Forgotten Books takes the uppermost care to preserve the entire content of the original book. However, this book has been generated from a scan of the original, and as such we cannot guarantee that it is free from errors or contains the full content of the original. But we try our best!

Truth may seem, but cannot be:
Beauty brag, but 'tis not she;
Truth and beauty buried be.

To this urn let those repair
That are either true or fair;
For these dead birds sigh a prayer.

Bacon

THE AMERICAN BOYS' BOOK
OF BUGS, BUTTERFLIES
AND BEETLES

The Trail Blazers Series

Boys of all ages from twelve to ninety are setting the seal of their approval upon these volumes. In addition to being stories of breathless adventure, each book pictures certain phases of American history which are not very well known. This background of history gives added pleasure and profit in the reading.

GOLD SEEKERS OF '49

By EDWIN L. SABIN. Illustrated in color and black and white by Chas. H. Stephens. 12mo. Cloth. $1.25 net. Postage extra.

BUFFALO BILL AND THE OVERLAND TRAIL

By EDWIN L. SABIN. Illustrated in color and black and white by Chas. H. Stephens. 12mo. Cloth. $1.25 net. Postage extra.

ON THE PLAINS WITH CUSTER

By EDWIN L. SABIN. Illustrated by Chas. H Stephens. Frontispiece in color. 12mo. Cloth. $1.25 net. Postage extra.

WITH CARSON AND FRÉMONT

By EDWIN L. SABIN. Illustrated. 12mo. Cloth. $1.25 net. Postage extra.

CAPTAIN JOHN SMITH

By C. H. FORBES-LINDSAY. Four illustrations in color. Cloth. $1.25 net. Postage extra.

DANIEL BOONE: Backwoodsman

By C. H. FORBES-LINDSAY. Illustrated. 12mo. Cloth. $1.25 net. Postage extra.

DAVID CROCKETT: Scout

By CHARLES FLETCHER ALLEN. Illustrated in color and black and white by Frank McKernan. 12mo. Cloth. $1.25 net. Postage extra.

J. B. LIPPINCOTT COMPANY
PUBLISHERS   PHILADELPHIA

THE AMERICAN BOYS' BOOK OF BUGS, BUTTERFLIES AND BEETLES

BY

DAN BEARD

FOUNDER OF THE FIRST BOY SCOUTS SOCIETY
AUTHOR OF "AMERICAN BOYS' HANDY BOOK," ETC.

WITH ILLUSTRATIONS BY THE AUTHOR

PHILADELPHIA AND LONDON
J. B. LIPPINCOTT COMPANY
1915

ACKNOWLEDGMENT

For the use and arrangement of insects
on colored plates we are indebted to the
American Museum of Natural History, and particularly to Dr. Frederick
Lucas and Dr. Frank Eugene Lutz, for their
sympathetic and generous aid in the work.

CONTENTS

vii

Contents

THE AMERICAN BOYS' BOOK OF BUGS, BUTTERFLIES AND BEETLES

FORE TALK

A FORE TALK ABOUT INSECTS, BUG-A-BOOS, BUG-BEARS,
BUG-HOUSES AND HUM-BUGS
HOW THE WRITER LEARNED THE LIFE HISTORY OF
BEETLES—HOW HE USED THEM FOR HORSES.

AMONG the little folk of this world known as the insects, we find almost as many traits of character as we do among the human beings. We have the idle insects, the industrious insects, the warlike insects, the robber insects, the dead-beat insects, the stupid insects and the intelligent insects. We also have among them the low, degraded insects, dirty insects, clean insects, the sluggish slow-moving insects, the bright lively insects, the useful insects and the beautiful insects; all of them are interesting, all of them in one way or another are of vast importance to man, and a study of their habits is not only a source of fun but it is also a most useful study. Besides which, boys, nature lovers live longer and happier lives than ordinary people!

1

Probably the first collection the reader will want to make will be butterflies, not because butterflies have more interesting lives or bodies, or even are the most beautiful, for some beetles rival the butterflies in beauty, but because butterflies are better advertisers than any of the rest of the insects. They display their beauties, attract the attention of the boys and even of the more stupid grown people. I say stupid grown people, because one boy of twelve who is alert and fond of nature will see and observe more things than the best-trained naturalist of thirty. A boy of twelve has not had his mind bothered by worldly things which dull the perception of a man, consequently the boy will see more, feel more, hear more, and smell more than the older person.

Not long since I was in the Smithsonian Institution at Washington in one of the private rooms not open to the general public and there I was shown drawer after drawer of butterflies, some of them so closely resembling each other that only a scientist could detect the points of difference, and enough of them to probably cover an acre or more of ground.

Few of my readers will want to make such a vast

collection as that at the Smithsonian Institution at Washington, and probably none of them ever will, for the collection at the National Capital is made up of the contributions of many, many collectors, but some of my readers may contribute to the collection at Washington or exchange specimens with the people at Washington, whom they will find ever ready to assist them in their work and encourage them in their study.

Do not be afraid of the big men at the head of our country's scientific department; they are all good fellows, they love the boys, especially the young naturalists, even better than they love their treasured collection of dried bugs, butterflies and beetles.

Every one who has read Mark Twain's works is familiar with tumble "bugs," which are not bugs at all, but beetles. As a rule, beetles are hard-shelled insects with their wings covered up with two neatly fitting lids which give them a back not unlike a turtle's. Every boy in the Southwest has enjoyed himself on a summer day watching a pair of tumble "bugs" roll their ball along the ground. Perhaps he has put a twig in their path and laughed to see the tumble "bugs" stop pushing the ball to

knowingly walk around and investigate to see what
chocked the ball so that it would not roll, work their
heads as if they were nodding with self-approval
for having discovered the trouble, then proceed to
roll the ball around the obstruction.

The scarab or sacred beetle of Egypt is nothing but a tumble "bug"; the old Egyptians, like
the boys of to-day, were wont to watch the tumble

"bugs" roll their ball along the ground; the
Egyptians thought they rolled this ball from sun-
rise to sunset; and because of the thirty joints in the
scarabs of their six feet they came to the conclusion
that these joints represented the thirty days of the
month. Then they set their imagination mill to
working and deified the tumble "bug." Even the
Roman soldiers wore a tumble "bug" on their sig-
net rings. Tumble "bugs" are funny, but people
are sometimes funnier than any bug.

There are some beetles so large that they would frighten timid people and some so small that one must use a magnifying glass to properly see them. They are all of them strong in proportion to their size; many of them are armed with pincers, like the well-known pinch "bug" of the Southwest, the friend and playmate of my youth. Not long since when I was travelling in the southwest, one of them flew into the car window and fell on the floor alongside of me, then reared up its familiar mahogany-colored body and opened its jaws ready to fight the whole world. I had not seen a live one since I was a boy and I felt like hugging the saucy little fighter. The vagrant poodle told of in "Tom Sawyer," came idling along the aisle of the church and sat on one of these same pinch "bugs." A pinch "bug" rightly administered can always create considerable excitement.

Besides tumble "bugs" and pinch "bugs" there are beetles of such brilliant colors that they look like jewels and people wear them set in brooches, stick-pins, sleeve-buttons and ear-rings. Some beetles carry lights at night on their shoulder-blades, others carry a lantern at the end of their jointed body, some are queer, some are funny, some are beautiful and some look dangerous, but I know of no beetle that carries a sting, and if you know how to pick them up, you need not fear their pincers. When you want to take up a live beetle, grasp it between the thumb and forefinger on each side of the division which marks the waist line, that is, the line which separates the shoulder piece, called the thorax, and the body piece to which

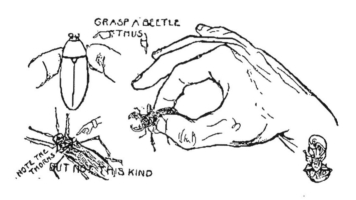

the wing covers are attached, and then it will not harm you, nor you it.

BUGS

In the good old days of our grandfathers it was the custom to quote from the Scripture, no matter what the subject of the discourse might be, and I might, if I had one of their old Bibles, head this fore talk with Verse 5 of Psalm XCI, which in the old translation from the Hebrew reads, " Thou shalt not be afraid of any bugs by night." But the Psalms were not referring to hemiptera, they were referring to the old meaning of bug as a frightful object of false terror, and in the later translations we find the same verse reads, " Thou shalt not be afraid of the terror by night." Possibly it would be even a better translation if it read, "Thou shalt not be afraid of any nightmare by night." You see, bugs then stood for some imaginary hobgoblins or terrible nightmare things which never had any existence out of dreamland. Thus we know a bug-bear to be a frightful goblin in the form of a bear, and a bug-a-boo a sort of nightmare creature which you are afraid is going to jump out and shout " boo " at you. The truth is, they were all hum-bugs.

In Wales they call a ghost a " bug "; among doctors and surgeons a bug is a tiny little terror, germ or microbe whose presence in one's system causes disease and death. One cuts a finger, gets blood poisoning, and the doctor, looking solemn and shaking his head, gives it a scientific name, but to his friend the other doctor he remarks, "He has a bug in that wound all right, and he is going to have a serious time of it."

There is nothing about the beautiful butterflies and big moths or even the beetles which need offend the sensibilities of the most squeamish and silly people; there is nothing creepy or uncanny about them, but one cannot truthfully say the same of the tribe of bugs. The very fact of their being called bugs tells us that they were looked upon as unpleasant things. Nevertheless the bugs are very important in this world and consequently are interesting creatures, hence a good collection of them is most valuable.

Many of the bugs are quite large, but although they are big, they are not the big bugs of human society, neither are they bug-a-boos, bug-bears or the inhabitants of bug-houses; they are the creatures naturalists call "Hemiptera." But from the foregoing you can see that the slang term " bug " for "bug-house" is only using the word with the old meaning of " bug " as a terror, as a nightmare; consequently it is very nearly correct to speak of a lunatic asylum as a "bug-house," in other words a " nightmare house," for if that term does not describe it, it will be difficult to find a better one in the dictionary.

But don't let this worry you. Not only do
nature lovers live longer than the ordinary people,
but they never go crazy and hence are in no danger
of being confined in the bug-house. But it is a
good thing to look up the meaning of these words,
because we all talk too carelessly. Suppose, for
instance, I should tell some English boys to collect

beetles, they would bring in a lot of cockroaches,
and if I should tell them to collect bugs, they would
bring in a most unpleasant collection of little creatures with which all travellers have been forced to
be altogether too familiar, and hence have little
desire to see a collection of them. But, if I should
tell the American boys to bring in a collection of
bugs, there would be nothing in the insect world

that they could capture which would not be found in their collection.

If you will go to the pumpkin vine, the gourd vine, the squash vine, you can probably find some of the ill-smelling insects known to the country boys as " stink bugs," and to the farmer as " squash bugs "—these are real bugs.

In the United States we have about sixteen hundred varieties of bugs which have been labelled, but Prof. R. P. Uhler, of Baltimore, is quoted by Mr. Leland O. Howard as saying "there are probably five thousand species of bugs in the United States and he thinks that fifty thousand would not be too large an estimate of the number of different bugs in the world." From this you may learn that if you want to get down to business and make a complete collection of bugs there will not be time for butterflies and beetles, nor will you have much time to devote to any other branch of study or play; still, one can make a fine collection without giving up all of one's time to it.

Bugs, like women, seem to be very fond of perfume, but, like some of the women, the perfume they use is not always the kind we would choose. The squash bug and the chinch bug have not

selected their perfume with the care we should wish;
some of the other bugs, though, as well as some
beetles, have the odor of ripe fruit, some smell like
cinnamon and spices, which is not so bad and a
little whiff is rather agreeable. The odor of bugs
comes from an easily evaporated (volatile) oil which
is hidden in the tubes of the body of the bug and
the creatures probably squeeze this oil out at their
pleasure and use it as a perfume, not always like
the ladies, to make themselves attractive, but sometimes apparently to make themselves so disagreeable that birds, toads and other creatures will refuse
to eat them. Like the skunks among the mammals,
the repulsive odor of some bugs seems to be their
gentle art of self-defense.

The big bugs among bugs, using the term as
they use it in society (not to represent the natural
size of bugs), are the true bugs, they belong to the
" 400." Nothwithstanding that they are the swells
of bugland, if some fatal plague should wipe out
all the bugs in creation there are not many of us
who would weep over their death, yet even this
event might in some unlooked-for manner upset the
balance of nature and cause disastrous results.

All bugs are " suckers," they have a long nose

like a bill, proboscis or beak which they poke into
the plant as does the squash bug and thus suck its
juice, poke into the skin of other bugs and caterpillars and suck out the juices, or poke into our
skins like some well-known bugs, and thus suck
our juice, or poke their bills into the openings of
bivalves (clams), into the bodies of snails and even
into the bodies of small fishes and suck their juices,

as do the large water bugs.

While I am dictating
this, there is in one of my
aquariums in front of me
a dead goldfish killed by
a water bug much smaller
than the fish.   These water
bugs are not always successful in their attempts
to suck the juices out of other creatures.   One
I kept in an aquarium thrust its long imperti-
nent nose into the shell of a fresh-water clam.
It was a small bivalve, about the size of one's
finger-nail, but when it felt that inquisitive nose
come into its private apartment it closed its little
doors tightly and quickly, and for three days that
water bug was forced to swim around with a clam

shell pinched on to the end of its proboscis and probably it had a sore nose for days thereafter.

My readers have a great advantage over the boys of yesterday—they have an advantage in the fact that they now have books written for them to tell them these things, also nature studies in all the schools, besides parents and teachers who are in-

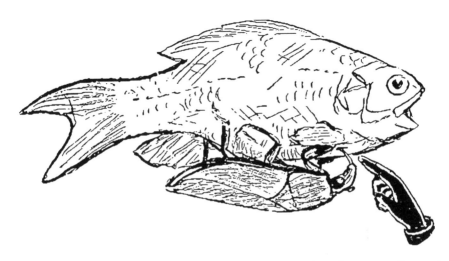

terested in such studies, whereas the boys of yester-day had no such books and the only nature stories printed were too absurd for a place outside of Mother Goose.

In spite of the dearth of books on our insect neighbors, however, when the writer was five years old he had learned by personal investigation the whole history of at least one beetle, he knew the male from the female beetle, he knew the eggs and nere they were deposited. He knew that the young were grub-worms that he used for bait when fishing for sun-fish, and he found all this out by watching the beetles themselves. He was very much puzzled when he watched the female grape-ne beetle deposit her eggs because the eggs were uch small objects compared to the beetles and there were *no little beetles;* they all seemed to be the same size, that is, all the light-colored female

beetles were of one size and the darker colored males a little smaller, but there were no baby sizes, no half-grown sizes.

The writer used these beetles for play horses, hitched them up to little paper sleighs, fed them on grape-vine leaves and kept what might be called "stables of them." There was another kind of beetle of a brilliant metallic green that he had frequently seen in the neighborhood of rotten stumps; this excited his curiosity and caused him to dig into

the decaying wood and bring to view many gru
worms; then he discovered some mummy-like creatures which were not grub-worms and not beetles,
and he also found some perfect and evidently brand-
new beetles. That set him to thinking and at last
it occurred to him that the grub-worms were the
baby beetles and the mummies were grub-worm
changing their forms.

The writer's mother had once shown him where
to hunt for the chrysalides of butterflies on the
under side of the top rail of the white paling fence,
and he had often found the pretty jewelled sleeping bag or chrysalis which covers the baby butterfly
while it is hanging head downward under the protecting rail, and he knew that this shell concealed
the caterpillar while it was changing form; hence,
a glance was sufficient for him to know that these
things he found in the rotten stump which were
neither grubs nor beetles, but helpless things half-
way between, corresponded with the chrysalis state
of the butterfly.

It was a grand discovery for him; he now knew
that the grub-worms were young beetles! He
shouted and danced with delight, for it was his
first real scientific discovery; no one had helped

him with the beetle problem and there was no one
except his good mother with whom he could share
his triumph because nobody else in those days
seemed to care whether beetles were born ready-
made or lived a baby's life as grub-worms. There
was no one but his mother to sympathize with him,
everybody else looked upon the studies of a country
boy simply as a sign of his being queer and uncanny;

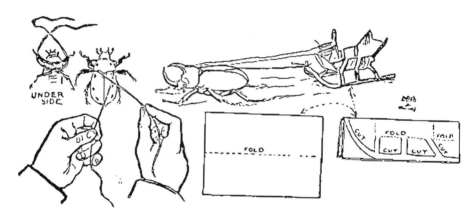

it seemed strange to them that a child should take
any interest in grub-worms! But this did not cool
his enthusiasm because he did not love nature for
the personal glory the knowledge of it would bring
him, and he did not study it to gain the approval
of the other boys; he loved nature because he could
not help it, the love was born in him and it is there
yet, and he is writing this book because he thinks
it is born in all children! Young people all pos-

sess it, although they may not know it, but as soon
as they find it out, the author believes they will
become as enthusiastic as he was himself when, as
a barefooted little urchin in northern Ohio he made
his first scientific investigation and discovered that
grub-worms were baby beetles.

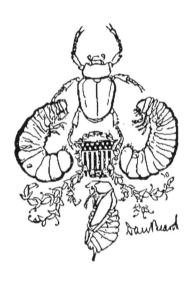

CHAPTER ONE

BUILDING A MAKE-BELIEVE INSECT
COMPARING A BEETLE WITH A BOY

IN order that we may understand the plan upon which insects are built, and, for that matter, the plan upon which every live creature is built, we must compare them to something we understand; probably the easiest way to do this is to pretend or make believe that we are about to create an insect ourselves, that we have in our hands some putty, clay, dough, chewing gum or modelling wax; the latter is best, so we will call it wax and from this stuff we are going to model the live creatures.

First we will roll the wax between our two hands (Fig. 1) and make of it a sort of worm, a kind of fat angle-worm, or, as the boys call it, a fish-worm. This, you will see, looks like a worm, and feels like a worm, but it is not alive and cannot move and if it should become alive it would not live long because we have made no provision for supplying new flesh and skin as the old ones wear out and waste away. To supply this need, we must have a mouth and stomach; in other words

18

our worm must be hollow all the way through from one end to the other so that it may take in fuel in the form of food to keep its engines going, absorb the good part of the food and throw the refuse or ashes away.

With a broom-straw (Fig. 2) we will punch a

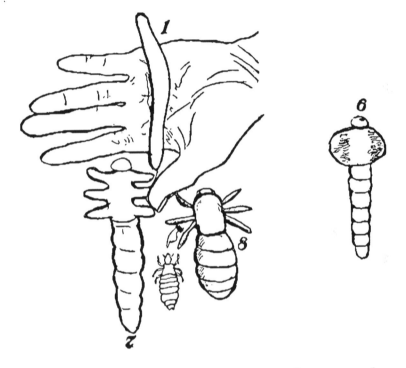

hole in our worm from end to end; now then, if some fairy will kindly arrange inside of this waxworm the proper tubes to soak up or absorb the good part of the food, then if this fairy will touch this thing with her wand and give it life it will be a very crude, but possible, form of a worm. It

would be hardly less crude, however, than that
strange creature we find just below low-tide mark,
called a " sea-squirt " by the boys and an Ascidian
by the school-teachers; but as this book is not about
worms or sea-squirters, we are not interested in
these things at present beyond the fact that we
begin with this form of life only because it is very
simple and easily understood.   In fact, it is so
simple that it would be hard for us to tell which is
the head and the tail of the wax-worm just made.

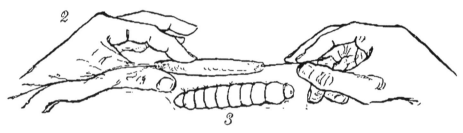

But do not let this worry you because our wax-
worm does not differ in this respect very greatly
from some forms of real live things.   In order to
make our wax-worm look like a caterpillar, we will
tie a number of threads about its body (Fig. 3).
The first section we will call its head, the next section, which we have made bigger than the head,
we will call its shoulders or chest and the other
sections we will call its body, belly or paunch.

We are making believe that the fairy has given

life to our wax-made worm and it can absorb food, but it has *no feeling,* it has no sight, no taste, so that it will eat any old thing as food. This is because we have not supplied it with the battery, so to speak, and connecting telegraph lines which we call nerves and which make it possible for live creatures to see, taste, smell, and feel. To do this it will be necessary for us to run a telegraph line through our wax form, from end to end, and to have small branch lines running to the surface. Fig. 4 shows one of these telegraphic systems such as is really found in a caterpillar. Now then, whenever these wires are short-circuited, our wax-worm will be doubled with pain. The principal difference between this system in the caterpillar and the system in the body of the reader lies in the fact that the central station is not of so much importance

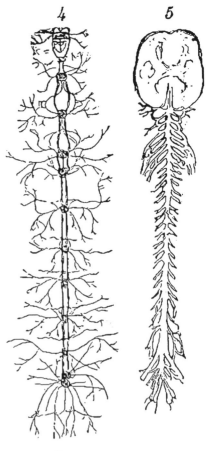

in the caterpillar as it is in the man—in truth, the caterpillar has numerous sub-stations and it might be said that it has a separate brain for each ring of its body. The stations are made by bunching a lot of wires together, that is a lot of the nerves, and making a " ganglion " of nerves which in Fig. 5 we call the brain. The lower forms of animals practically have many brains or one brain running from one end of their body to the other, so that when you cut the creature in two pieces each piece is alive and remains alive for some time; but if you cut a man in half, or cut his head off, you sever the cable, that is, you disconnect the wires, the spinal cord, and all feeling ceases,—in other words, he is dead. Fig. 5 shows a rough plan of your own telegraph system with the central station at the top. Of course there are branch nerves which run off to your arms, legs and all parts of your body, but these have been omitted and the diagram simply shows the main cable lines.

Besides having a telegraphic communication in your body like that of a caterpillar, you also have the hole punched through it which you call your mouth, throat, etc.

But we must not forget the wax on which we

are at work; of course it should have some legs. These we will make by pinching and flattening the sides of the first joints behind the head * (Fig. 6), after which we will cut the flattened side into six flaps (Fig. 7); next we will roll these flaps between our fingers and make legs of them, then we push the tail towards the head, thus crowding the rings together in the form shown by Fig. 8. Our wax thing now begins to look like an insect. A very low and degraded form, it is true, but we must have a creature with a hole for its mouth and a tube for its stomach and six legs with which to walk. Most insects, however, are supplied with wings of some sort and these may be easily made; we have, however, gone far enough to understand, in a general manner, the construction of the little creatures about which we are to talk through the rest of the chapters of the book.

Of course you know that every live thing which is not a plant is an animal. A beetle, a worm, a fly, a bug are animals. The creatures you generally call animals, such as dogs, cats, horses and elephants, are animals, too, but they belong to the family of milk-givers called mammals. But bugs,

---

*See illustration, page 19.

butterflies and beetles are not mammals, they are
not warm-blooded milk-givers. In order that we
may be certain that we understand this subject and
at the risk of being thought undignified, I have
drawn some pictures here of a boy and an insect,
showing their similarity and their difference.

As we have already suggested, the best way
to understand anything which is mysterious to us
is to compare it with something with which we are
familiar. Now, then, so that we may not scare the
reader with a dull talk on comparative anatomy,
we will skip all the big words and get down to
what the boys in their slang talk call "brass tacks,"
which if I understand aright means "bottom
facts."

To begin with, we know, of course, that the
reader does not look like a beetle, bug or butterfly,
but we also know that there are certain things which
all live creatures possess in common. All live creatures must have blood or some sort of juice which
serves as blood, all live creatures must have a head
and some sort of breathing apparatus. All live
creatures must have some kind of a hole for a
mouth, something which acts as jaws, teeth, tongue,
throat and stomach. Also most live creatures must

have some means of locomotion, that is, moving
from one spot to another; in the higher orders of
life these organs of locomotion are called "legs."

But one of the first differences which any child
will see between beetles and himself is that the
former creatures have a skeleton on the outside of
their bodies with their muscles and blood-vessels

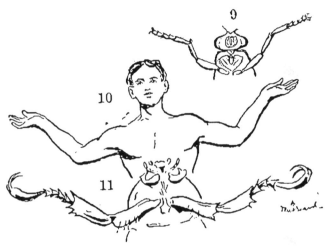

and all their internal organs located *inside their
bones,* while with himself the reader knows that
the muscles, blood-vessels, nerves and other organs
are plastered, so to speak, on or around the framework of his skeleton. In other words, the human
skeleton is the framework of the body like the
frame of a kite, the framework of a boat or the
framework of a house, and our own frame or
skeleton's use is evidently to stiffen and to hold our

body together and keep it from sagging down into a helpless lump like a bag of meal.

By referring to the diagram (Fig. 12) the reader will also see that the insects have six legs instead of four. We have four legs, our front or fore-legs we call arms, the hind ones we call our legs. But the fly (Fig. 9) and the beetle (Figs. 11 and 12) and all other such creatures not only have

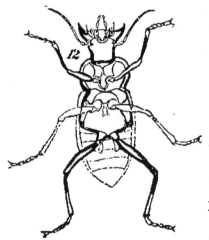

legs and arms like a human being but they also have a middle pair of legs.

In the illustrations (Figs. 9, 10 and 11) are shown the head, arms and chest of a man, also of a common house-fly (Fig. 9) and just below the man that of a spotted yellow grape-vine beetle (Fig. 11). Roughly speaking, there is some resemblance between the three—each has a head, a body and front legs or arms. The head of the fly and the head of the man are separated from the chest by a more or less slender neck, but the beetle's head is jammed into its chest. Following these three diagrams is one of another beetle (Fig. 12)

the scientific name of which is Harpalus caligino-
sus—pardon the big name, I did not intend to use it
—but the name has nothing to do with the diagram,
which shows the front side of the beetle, that is,
what would be the front side of the beetle if the
latter walked on its hind legs like a man; in reality
it is the under side of the beetle.   In the diagram
(Fig. 12) I have shown by
dotted lines the parts which
do not resemble the man,
that is, the extra pair of legs
and its belly, which it carries on
its back, and in the diagram of
the man (Fig. 14) is shown,
with dotted lines, the outside
covering of the bones of the legs
and arms, for to make the man
like the beetle we must strip off the outside covering
of muscles from the bones and put muscles and
blood-vessels and nerves inside of them.

Besides the diagram of the man (Fig. 14) is
the rough chart showing the muscles on a man's
leg, also an insect's leg split in half so that one
may see the muscles on the *inside* of an insect's
leg (Fig. 13).

We will not carry this comparison any further because this book is about bugs, butterflies and beetles and not comparative anatomy, but it is important that you understand in a rough way just how like or, if you choose, unlike, we are to these tiny creatures.

To make an insect of a man, you would have to prolong the man's body way down below his knees,  push his skeleton out in the front, give him another pair of legs, cover his back with a shell like a turtle and make him creep on his feet with the front side of his body next to the ground. There are other things you would have to do with his back. You would have to arrange for wings; in fact you would have to do so much to the man to make an insect of him that the job would not be worth while, besides which it would not be exactly proper to so treat a man, because according to many scientists it has taken centuries and centuries for man to evolve, that is, to grow from some sort of pulp or jelly-fish to a land animal, to a missing link, and then to a man, and it would be unkind to send him away back to the insect world.

We have no exact record of man's growth to his present dignified position in nature, but every one of my readers can see the transformation of an insect, see how it grows from an egg to a worm, from a worm to an animated mummy, from a mummy to a perfect winged beetle, beautiful moth, or gorgeous butterfly according to the particular kind of eggs first observed.

CHAPTER TWO

HOW TO EQUIP ONESELF FOR COLLECTING INSECTS
HOW TO IMPROVISE BOTTLES FOR ALCOHOLIC SPECIMENS
HOW TO HAVE POISON BOTTLES MADE
HOW TO MAKE DRYING BOARDS AND SPECIMEN BOXES
HOW TO MAKE BUTTERFLY NETS AND HOW TO USE THEM

THERE is no doubt, boys, that you are of great importance on this earth, and in all my writings and in all my talks I have taken pains to show how important you are. But do not be conceited; if there is any danger of you thinking you are " IT," to use another one of your expressions, you have " another think coming," for you are only living on this earth by permission of the birds. If all the birds were killed, the insects would eat up everything in sight, they would devour the forests, and the world would be an uninhabitable desert.

It is an exceedingly dangerous thing to upset the balance of nature or, as my good friend Doctor Hornaday puts it, " to monkey with nature's buzzsaw." Bugs, butterflies and beetles are a busy lot, they need watching, they are mischievous little gnomes, but the Great Creator supplied the earth with birds to keep these little insect fairies in subjection. Why, one pair of gypsy moths, if left alone, under favorable conditions, can produce enough caterpillars in eight years to destroy every green leaf in the United States; the Kaiser, the Allies and all the guns, aëroplanes and submarines could not possibly do as much damage as one pair of gypsy moths and their children.

Suppose there were no birds, and the little bug called the hop aphis, which infests the hop vine, were left alone and unmolested. After a careful calculation, one naturalist tells us, and we have no reason to doubt what he says, that a pair of these little hop bugs would breed so fast that in less than a year there would be six sextillions, 6,000,-000,000,000,000,000,000 children, grandchildren, great-grandchildren and great-great-grandchildren, etc., all sucking the juice out of the hop vines. This number is too big for us to get the proper perspective view of it, unless we put it another way, and a Mr. Forbush has done this for us. He has figured it out about this way: If you place these little bugs, ten of them to an inch, on a straight line, then shoot the line of them up into the sky, it will reach so far into space that, should the last little aphis on the line flash a light as big as

the sun, it would take TWO THOUSAND FIVE HUN-
DRED YEARS for the light to reach the earth! Think
of it! If this line had been projected up into the
sky, into space, and the last little bug had flashed
his light five hundred years before the birth of
Christ, we, to-day, would not yet have been able to
see that light, we would not even know it was there.

Creeping, crawling, flying, burrowing over and
under the crust of this old earth, are about one
million different kinds of creatures, of which we
have only labelled and sorted out about three hundred thousand varieties. Of course you will not
find them all in this book or any other book; it
would take many of Carnegie's largest libraries to
hold enough books to describe them all.

The writer has lived quite a while, but during
his lifetime there have been only a few mammals
discovered (you see the mammals or milk-giving
animals are so big that they are easily found if one
visits their haunts), but every day we can walk over
new bugs, butterflies and beetles without seeing
them, or miss them even when hunting for them;
this makes the game fascinating and much more
interesting and useful than collecting birds' eggs
or birds. Why, every bird wears a halo around

its head if you could only see it, and it is worse than a crime to kill them.

I am telling you all this to impress upon your minds the importance of your work and play in collecting and studying insects and because there are a lot of good-hearted, sentimental women who do not use their heads to think, and consequently tell you that it is cruel to collect butterflies, to collect beetles and to kill caterpillars, which is not true; but it *is* cruel, mean and selfish to destroy the birds.

If you intend to make a collection of bugs, butterflies and beetles, begin by first making a collection of small bottles such as are used to contain homœopathic pills (Fig. 16) or the sort sometimes used to hold individual fancy cigars, also any other small wide-mouthed bottles which you can procure. Making this collection in itself will be fun. While you are doing this, it will not be amiss to make a collection of all the corks you can get hold of; they are just the things you want to which to pin insects. Then make a collection of small tin boxes used to hold small quantities of tobacco and also those used to contain some kinds of preserved foods sold in the grocery and delicatessen stores,—speaking about delicatessen stores reminds me that some firms also put up things in glass jars about the size of a half tumbler which make splendid bowls in which to hold water bugs, caddice worms and other creatures found in brooks or ponds. In the illustration (Figs. 15 and 17) are shown some of these different sorts of glasses which were this minute secured from the top shelf of the pantry.

The shortest one came from the delicatessen store, and if I remember aright originally contained some sort of preserved fish, but what was its original use is of no importance to us except as it suggests where to look for it.

If the boys of to-day, however, are anything like the boys of yesterday, they will be able to get a supply of these bottles, jars, tumblers and so forth without much trouble. In most households these things are thrown away after their contents have been used, and every ash dump has a supply of them.

Ordinary bottles with narrow necks are not good for live specimens as they do not supply enough air, while for both live and dead specimens they are awkward to handle because of the narrow necks and consequent danger of injuring the insect while introducing it into the bottle or taking it out. At the ten-cent stores I have been able to secure a number of small fish globes which are used by me in which to keep live water beetles, water bugs, skaters, boat-beetles, the larvæ, that is the young, of the dragon-flies, as well as snails, periwinkles and small fresh-water clams. The latter creatures are the food supply for the water bugs.

THE USE OF PILL BOXES

A lot of little wooden pill boxes are very handy for delicate or minute specimens, and it is a good idea to have cotton in some of the boxes on which to place your trophies.

PINS

The collector will need pins, but it is not necessary to buy the long German skewers, although they are made for this purpose. The German pins come in several sizes and are longer than our ordinary pins but not so thick. Professional bug hunters or, to use their chosen name, entomologists, use the professional pin and No. 1 can be used for minute specimens, little teeny-weeny bugs, gnats and so forth, but even then it is sometimes necessary to gum the little creatures on a piece of paper, so small are they, and then run the pin through the paper. If these German insect pins are out of your reach, use fine needles or even broom straws for your small insects and ordinary pins for the others.

EQUIPMENT

The next thing necessary in the preparation for your campaign as a collector, is to make drying boards (Figs. 19–24). When the writer was a small boy, he made drying boards for himself, and no doubt his readers can do the same. In a pinch, a stiff piece of writing paper (Fig. 19) may be pressed into service as a drying board.

When everything is ready to receive the captives, you must prepare some nets (Fig. 27) with which to catch the butterflies, grasshoppers and

flying creatures and some slumber bottles in which
to drop the captives where they will be over-
come with chloroform and other poisonous fumes
(Fig. 18).

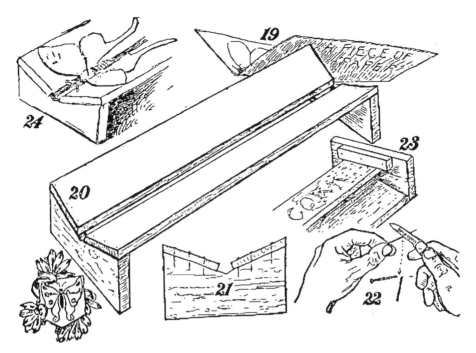

POISON BOTTLES

I have tried burning matches, I have tried
mashed-up peach-tree leaves, kerosene and camphor, but none of these makeshifts kill quickly
enough, they all give the victim time to flap around
and spoil itself as a specimen, so I think you will
have to spend a few pennies possibly for chloroform.

CYANIDE BOTTLE

Mr. H. S. Surface, M.S., of the Pennsylvania Department of Agriculture, advises the dropping of a lump of cyanide of potassium the size of a small hickory nut into the bottom of an empty bottle and covering it with dry plaster of Paris, after which he tells us to pour enough water on the plaster of Paris to make it set as you do cement. The proper way to dry this bottle is to set it upside down and allow it to drain until the plaster hardens. Next cut out a piece of blotting paper just the right size to fit over the plaster of Paris, like a gun wad over a charge of powder. It should be large enough to make it necessary to use force to crowd it down on the plaster, where it will then stay as a protection both to the insects and the plaster (Fig. 18). A slumber bottle or poison bottle of this kind must be kept tightly corked at all times except when the cork is momentarily removed in order to drop an insect into the bottle. A cream bottle makes a good slumber chamber. Of course, any boy with common sense will know better than to put his own nose over a bottle full of fumes poisonous enough to kill insects. To say the least, the breathing of these fumes will do him little good. Such a bottle should be guarded with care, for if it is broken, children might get hold of the contents with most serious results. Tell the druggist how to make a slumber bottle and let him prepare it for you. It is best to dissolve the cyanide first by pouring in the bottle enough water for the purpose, then sprinkling the plaster of Paris over the mixture until there is enough plaster to harden into a firm shell of cement. The druggist will not sell you cyanide unless you have a permit, as it is a dangerous poison; for that reason chloroform is still used by many to kill the insects.

THE CHLOROFORM BOTTLE

First put a wad of absorbent cotton in the empty bottle, then saturate the cotton with chloroform, and over this place a pad of blotting paper as already described for the cyanide bottle. The chloroform bottle, too, must be kept tightly corked or the chloroform will evaporate. All such bottles should be labelled with the skull and crossbones and the word POISON!

DANGER

Since the cyanide of potassium is sealed in the bottle with the plaster of Paris and further protected by the wad of blotting paper, there is practically no danger of any foolish persons injuring themselves with it, and as the chloroform is all in the cotton and also sealed by a wad of paper, there is practically no danger from it. But poisons, like fire-arms, are made to kill, and neither poison nor fire-arms *have any brains of their own;* they have but one duty to perform and that is to kill; you must supply the brains for them in order that they do no damage to valuable animals and human beings.

A boob who points a loaded or *unloaded gun* at anyone should be soundly thrashed for the act, and a boob who fools with poison and does not use the proper precautions in handling it should be treated in the same manner by any person who detects him in his carelessness.

HOW TO MAKE THE DRYING BOARD

There are two ways of drying a butterfly: one is with the wings perfectly horizontal, and the other is with the wings tipped at a slight angle. The position of the wings depends upon the slant of the side-boards (Fig. 20). To make the wings horizontal, that is, on a level with each other, the end board (Fig. 21) need not be notched or cut in on the bias, but the top of it may be level with the bottom, otherwise the drying boards are made in the same manner as the one shown in the illustration.

Of course the drying boards for big fat moths or night butterflies should have a wider slot than the one for day butterflies, which have narrow or slim bodies. In order that the reader may decide for himself, it would be best for him to go out into the fields and collect a number of butterflies and some big moths, like the one shown in Fig. 24, and then make the slots in his drying boards correspond to the size of the bodies of the insects.

First he takes the two ends (Fig. 21), cuts them exactly alike, so that when laid one on top of the other they both agree edge for edge with no overlapping. Next he takes two smooth pieces of soft pine wood, each exactly the same size as the other, for side-boards like those shown in Fig. 20; these he tacks on the end boards as shown in Fig. 21, using the little brad nails from a cigar box; or if he has no cigar box, he takes some ordinary pins and files off the points as shown in Fig. 22, thus making suitable brads for the purpose.

After the boards are put together, as shown in Fig. 20, he tacks a strip of cork under the slot, (Fig. 23), or he puts cleats on the end boards as shown in Fig. 23 (showing the under side of the drying boards), fastens the cork to them or puts the cork on in any way his ingenuity suggests, but it is necessary that it shall be firm and not sag.

COLLECTING NETS

There are numerous kinds of nets used for collecting water insects, but every net is more or

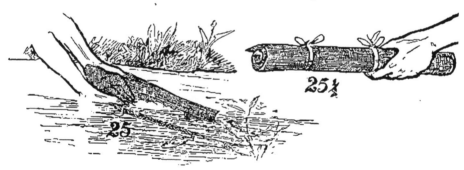

less awkward to carry on a hike and I have found that or the best things with which to catch aquatic (water) creatures is simply a piece of wire netting such as is used to screen the windows of our houses to keep out flies and mosquitoes. The piece I have is 17 inches wide and two feet long. I roll it up as shown in Fig. 25½ and in this position t is easily carried. When I want to use it, I unroll it, grasp each side of the piece (Fig. 25) and use it as a scoop, poking it along under the water plants until it is covered with duck weed, frog slime, pieces of water-cress, etc. Then quickly and carefully lift it from the water and dump the contents into a tin pail, or spread the wire screen out on a board and carefully go through the mess with the fingers, picking out the small creatures and placing them in vials or boxes. But I find the best way is to dump the whole mass into the pail and then do the sorting and hunting after I reach home. With a scoop of wire netting I can catch little fish, sticklebacks, snails, periwinkles, minute fresh-water clams and all the interesting and curious creatures upon which water-bugs and beetles usually feed.

BUTTERFLY NET RING

For land winged creatures, such as grasshoppers, katydids, devil's darning needles, moths and butterflies, we need an insect net (Fig. 27). To make this, take a piece of telegraph wire, bend it around and make a circle about a foot in diameter which, you know, means across through the centre from one side to the other. The two ends of the wire should be hammered on an anvil, or you may use a flatiron as a substitute anvil, until they are bent like the ones shown in Fig. 26. These two ends can then be forced into a stick of bamboo such

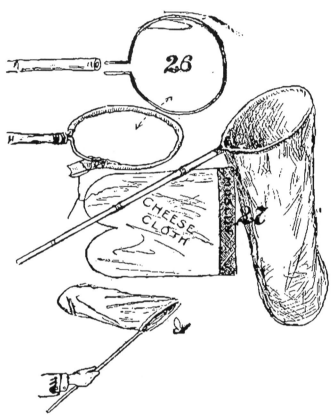

How to make and use a butterfly net.

as is used for a fishing rod, after which the end should be bound with bicycle tape, copper wire or twine to prevent the bamboo from splitting. The handle may also be made from a small broom handle or an old walking cane by neatly cutting two grooves one on each side of the stick, the length of the two ends of the wire, then placing the ends of the wire in these grooves and securing them there with a piece of bicycle tape or twine as already described. After this, a piece of muslin or an old piece of sheeting may be used to cover the wire and sewed there (Fig. 26).

THE NET BAG OR POKE

You may make a net of cheesecloth, mosquito netting or bolting silk such as is used in flour mills, or tarlatan, although this is usually too stiff and does not work as well as the foregoing, or a thin, light quality of swiss. What you need is a light, finely meshed but transparent cloth, one that allows the air to pass through it when the net is in motion, and allows you to see your captive inside of it after a successful sweep. The bottom of the bag or poke should be rounded as shown by the pattern in Fig. 27. It is well to sew a band of muslin at the top of your light material which you can stitch to your hoop and thus make your net stronger and less liable to tear. The net should be considerably longer than it is wide, about the proportion shown in Fig. 27. When you have captured a butterfly

(Fig. 29) be careful not to bruise and injure it, but as soon as its wings are folded together, as shown in the diagram, grasp the thorax, that is, the part of the insect corresponding to your chest, between the thumb and forefinger; do not reach into the net to do this, but grasp it from the outside and give the insect a pinch; this will kill it without disfiguring it (Fig. 29).

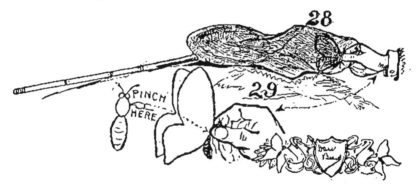

How to kill an insect by pinching it.

SPECIMEN BOX

You remember that you were told to make a collection of corks? Fig. 30 shows you a specimen box in which these corks are used, furnishing foundations upon which to pin the insects (Fig. 31); a cigar box or any sort of shallow box will do if it has a lid to it to protect its contents. To make a specimen box, take a neat clean piece of white cardboard (Fig. 32), cut out the corners so that you can bend back the edges and make the cardboard fit exactly in the box as shown in Figure 30. The advantage of this box is this: the space under the cardboard may be filled with camphor gum, moth balls or any other material abhorred by

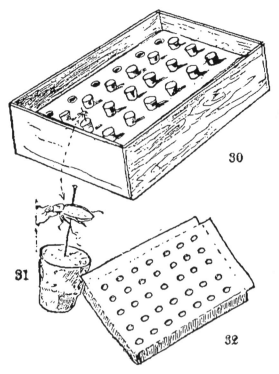

Specimen box with corks.

live insects, and after the corks are in place there will be no danger of the camphor or moth balls jolting around and injuring your collection. The cardboard (Fig. 32) is glued by its folded edges to the sides of the box (Fig. 30). As these edges are turned down, they do not show and it gives the box a very neat appearance. Another advantage of this specimen box lies in the fact that although some of the corks may be big and others small, you can cut out the holes to fit the individual corks and allow them all to be the same height above the cardboard and thus give a neat and uniform appearance. The ordinary way to make a specimen box is to line it with sheet cork; this is more ex-

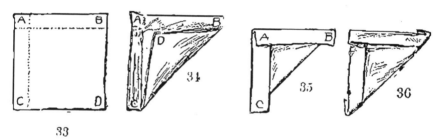

Folding paper for butterfly specimen.

pensive and to my mind not as convenient as the one here described, but corrugated brown paper, such as is used for protecting books and other merchandise when sent by express or mail, costs nothing and the box may be lined with it (Fig. 42).

BUTTERFLY ENVELOPES

These may be made by folding pieces of paper into three-cornered envelopes (Figs. 33, 34, 35 and 36) but I usually use the envelopes made for letters

like the one shown in Fig. 37. I fold one corner
to the centre as shown by the crease in Fig. 38, then
bend the other half over as
shown in Fig. 39, after which
I bend the flap F (Fig. 39)
over it as shown by Fig. 40.

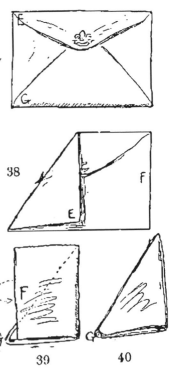

If the butterfly is care-
fully placed in the triangular
envelope and then put in a
box carried in your pocket
for that purpose, it will be
safe from injury until you
reach home. If it is too dry
for the spreading board,
place it on some wet sand in
a box and the moisture will
soften it and make it pliable.

How to use an envelope for a
butterfly.

ALCOHOLIC SPECIMENS

These are not the sort which you can see standing in front of the bar-rooms and saloons, although
they too are almost as soft-bodied creatures as the
caterpillars, grub-worms and spiders which we are
talking about and which we want to preserve, and
here is where all our homœopathic pill vials come
into play. Fig. 41 shows an ordinary pasteboard

4

box with holes cut in the cover for specimen
bottles. Fig. 41 shows also a block of wood with
holes bored in it and an arrangement of screw eyes,
also a grooved piece of wood with a cardboard
top to it. Fig. 42 shows cross-section of specimen
box with corrugated paper bottom and cardboard

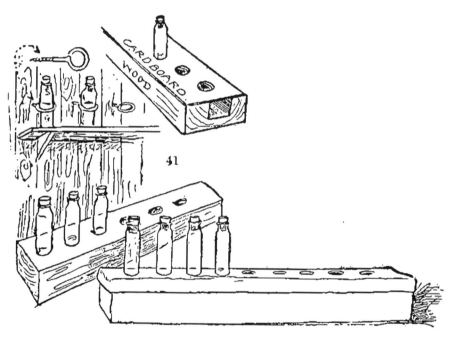

41

false bottom through which the pins are stuck. The
first pin shows how a minute specimen is gummed
to a piece of paper; this is a German pin made for
the purpose, the second one is a broom straw put
through holes punctured by a pin.

Fig. 43 shows how to extend and hold in
position the wings of butterflies and moths by the

use of strips of paper and any sort of pins. The
pins are not thrust through the wings, but through
the paper outside of the borders of the wings; this
must be done carefully so as not to rub the scales
off the wings and thus spoil the specimen.

Fig. 44 shows how to make a useful little tool

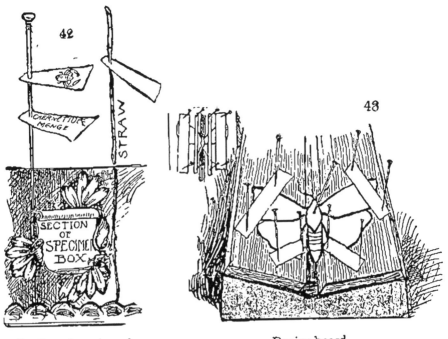

Section of specimen box.   Drying board.

by taking a small stick and inserting the heads of
two small needles in the stick, leaving the points
exposed as in the lower figure; these may be used
to spread the legs of a beetle. A stick with one
needle in it makes a useful tool in arranging and
handling small specimens.

There is always a danger of bending pins when they are stuck in the board by hand, hence pliers are usually used to grasp the pin in place of one's fingers.

It is not the object of the writer to tell the ways and means of manufacturing all the things you need, but it is his object to start you on your

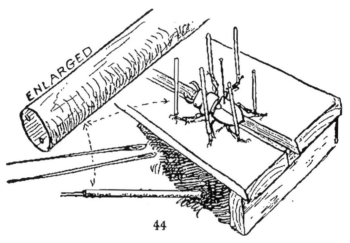

44

Drying board and needles for holding legs.

career with a few simply made contrivances and with the idea that as a good American boy with pioneer ancestors you have inherited the ability to think and devise these things yourself; if you are not an American boy, but come of foreign parentage, you have the United States History to go by, which is the history of your adopted country and tells about those old pioneers who are your ancestors by adoption, so that you must inherit their gumption, self-reliance and initiative, and inherit it by *adoption*. But if this book should fall into the hands of some nice little boys who have never whittled a stick or made a kite, they may be consoled with the fact that all the material necessary for collecting insects and preserving specimens may be purchased from firms dealing in and making a specialty of such merchandise, and it may be added, so can the specimens themselves, but what *real* boys want to buy specimens? We are out for the fun of collecting them, for the hike across country, for the exploring of the ponds and streams and scouting among the hedges!

CHAPTER THREE

THE BUTTERFLY AND MOTH FAMILY

To the young nature student it often seems as if the old naturalists and professors who write books lie awake nights to think of difficulties which they may put in the path of the amateur. They rummage among their Latin and Greek dictionaries to find long and impossible names to hitch on to the tiniest and smallest of creatures, names which no small boy may pronounce and which no big boy loves.

But do not think too ill of the old scientists—they are good fellows at heart and they mean well. You see they could not take the names which you use for things because the boy in another State uses different names for the same things. For instance, the fish called a " bass " up north here, is called a " trout " down south, the bird we call a " bob-white " is called a quail in Ohio and a partridge down south, while the ruffed grouse is also called a partridge and a pheasant.

This way of mixing things up drives naturalists to hard names; besides, if they should use

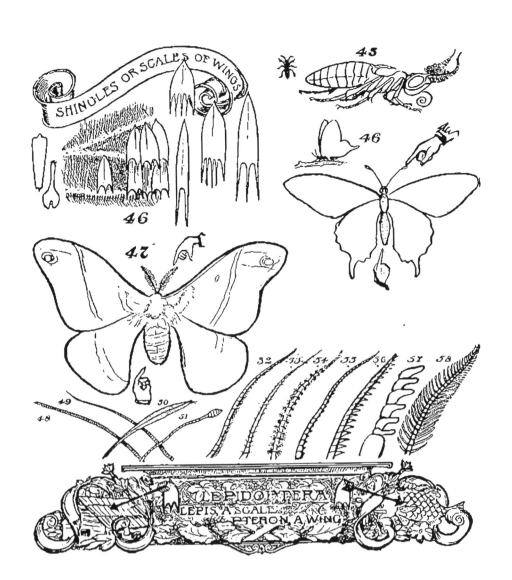

SHINGLES OR SCALES OF WINGS

43

46

46

47

49

48

30

51

32 33 54 53 56 57 58

LEPIDOPTERA
LEPIS A SCALE
PTERON A WING

French, German, English, Italian or Russian names, it would make everyone angry who did not speak that particular language as their native tongue.

But when it comes to Latin and Greek, these languages are so dead that they are dried up like Egyptian mummies and are only used by scholars, priests and scientists, and from these languages naturalists select their names for bugs, butterflies and beetles, with no one but the boy to object. Hence they call the moths and butterflies Lepidoptera, making the word from *lepis*, a scale, and *pteron*, a wing—in other words, a scale-wing.

If you will rub your finger-tips across the wing of a butterfly or a moth, the velvety surface of the wing will come off and stick to the ends of your fingers, and when you examine this dust with a powerful magnifying glass you will see that it is composed of very small scales shaped like some of those shown in Fig. 46.

The Scale-wings are divided into two families, one known as the butterflies that fly by day and the other as the butterflies that fly by night, or as

butterflies and moths. But when we speak of
moths, you must not imagine that they are all as
small as those tiny ones whose babies feed upon
our woollen clothes and furs when the latter are not
properly packed away for the summer. Some
moths are very large indeed and very beautiful;
both butterflies and moths have six legs (Fig. 45)
and four wings (Figs. 46½ and 47) and a pair of
feelers or smellers (antennæ) (Figs. 46–58).

As you can see by the diagram, and as you
know by looking at the live insects, the wings of
the butterflies and moths are, as a rule, very broad
and are shingled with minute scales. The wing
itself is a thin paper-like skin which is stiffened
by a framework of branching ribs or veins (Fig.
59). These veins may easily be seen when the
wings have been rubbed between one's fingers.

The lepidoptera have small heads and a tongue
rolled up like a watch spring under their face;
they can uncoil their tongues when they want to
insert them into flowers to reach the honey concealed there. They use their long tongues in much
the same manner that you use a straw in a glass
of lemonade.

It is not the butterflies and moths which do

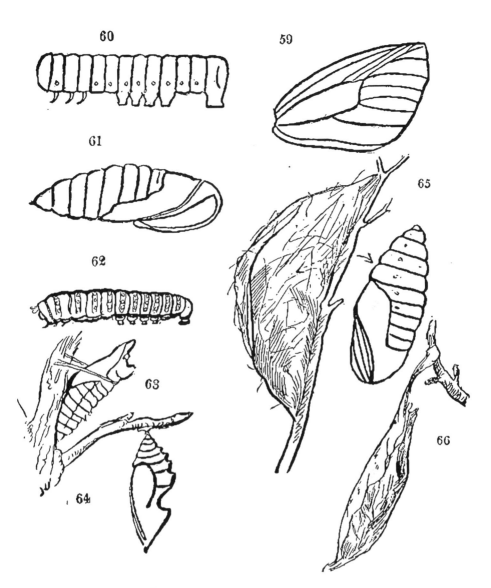

FIG. 59. VEINS ON WING.
FIG. 60. DIAGRAM OF PARTS OF CATERPILLAR.
FIGS. 61, 63, 64. PUPÆ OR MUMMY.
FIG. 62. A CATERPILLAR.
FIG. 65. A COCOON AND CHRYSALIS.
FIG. 66. A COCOON.

the harm in this world, but it is their children, the caterpillars (Figs. 60 and 62). It is a baby moth that eats our woollen clothes and furs and it is the babies of the bigger moths and butterflies which eat up our garden truck, play havoc on the farm and with the forest trees. The mother butterfly lays its eggs on the plant which its babies are to use for food, the eggs hatch out into tiny caterpillars, these caterpillars do nothing but eat! eat! eat!! When they grow too big for their skin, a new skin is formed underneath the old one and they crack open the old one and crawl out to eat some more.

They keep this up until they begin to feel queer, then they know it is time to stop eating, something mysterious is going on inside of them and they are about to change to " pupæ." This is a word which means something wrapped up in swaddling clothes (Fig. 61, 63, 64 and 65) like an Indian pappoose.

This pupa shape is formed inside the skin of the caterpillar and it wiggles its way out through the caterpillar's skin. The butterfly pupa we call a chrysalis, the pupa of a moth is usually concealed inside of a silken cocoon (Figs. 65 and 66) or in a little cell or cave underground. After a while the skin of the pupa or chrysalis cracks, and out crawls a limp, damp, flabby looking creature. For a while this limp object spends its time trembling and shaking as if it had the ague, but it is really shaking the wrinkles out of its crumpled wings and allowing the blood and juices to circulate through the veins and ribs of the wings until they are fully expanded like the paper stretched upon the frame of a kite, then the soft veins and ribs in the wings harden and stiffen and the perfect butterfly or moth is ready to fly.

BUTTERFLIES

The butterflies which you usually see have slender bodies (Fig. 46½), and when they are at rest they will fold or close their wings as one closes a book, bringing them together and holding them upright; also they will probably own clubbed feelers or antennæ (Figs. 46½ and 51), whereas the ordinary big moths that you meet will probably have fat bodies and feathered antennæ (Fig. 47).

It will surprise you to learn that our beautiful moths and butterflies belong to a lower family than the hymenoptera—this is another one of those big words which is made of *hymen,* meaning a skin, and *pteron,* a wing, skin-winged insects. These are the bees, wasps, ants and saw-flies. The bodies of the butterflies and moths are soft, while those of the bees, wasps and ants are hard and more like armor.

The butterflies' wings are very big compared to the hymenoptera and their mouths are especially made for them, a style of their own, what naturalists would call "highly specialized."

The young butterflies are worm-like babies (larvæ) and all these things, according to naturalists, go to show that our gaudily dressed idle butterflies do not move in the same circle of high-brows as do the bees and ants, that they are not of as good a family. Their legs are little used, the arms or fore-legs of some butterflies being little more than ornaments or decorations to their body and almost as useless as the buttons on a man's coat sleeve or at the back of his frock coat.

Our lepidoptera, our moths and butterflies, are essentially airmen and not hikers; even with a big handicap in their favor, the laziest ant would leave them far behind on a hike.

When you go into the business of caterpillar

farming and raise a lot of these greedy creatures, you will find that they often show a high order of insect sense, but the caterpillars seem to leave all the sense they have in their chrysalides and the butterflies themselves are not remarkable for their brains—not nearly as remarkable as they are for their beauty; brains and beauty do not seem always to go together.

But when you start to capture a butterfly and to pursue him across the fields, you will find that its seemingly aimless flight is not so aimless as it appears. The butterfly is using the same tactics and for the same reason that a big armored cruiser does whenever the outlook spies the periscope of a submarine poking up above the waves. Many a time I have been outwitted by a butterfly which I thought would be easy to capture. Still, they have not the brains of the wasps, bees and ants.

" The ebullition of voluntary energy of the larvæ is sometimes remarkable; " but " they are rarely footless, usually possessing from one to five pairs of embonpoint, abdominal props, besides three pair of corneous jointed thoracic limbs! " That's the way some of our teachers would speak of a caterpillar, for it is much easier to spill these

words over a page than it is to find simple ones to
tell the same story.

Nevertheless, boys, we are going to stick right
to the talk that we can understand as closely as
the subject will allow us. To swallow two such
words as Heterocera and Rhopalocera right on
top of Lepidoptera would give any boy indigestion
of the brain and a pain in his mental tum-tum which
would unfit him for butterfly hunting and make
him dream that he had corneous jointed limbs on
his abdomen, and could never again slide down
hill belly-buster.

CHAPTER FOUR

AMERICAN SILK-WORMS AND GIANT NIGHT-BUTTERFLIES, MOTHS, OR MILLERS

No matter how careful naturalists may be to explain what a moth is, all people who are not naturalists will continue to think that there is only one kind of moth, the kind which eats up clothes. Night-butterflies is too long a term, but the children's name of miller is short, easily remembered, and generally understood; besides, the insect is called a miller because it is apparently covered with dust. So we shall adopt that term. Fig. 67 shows the caterpillar, Fig. 68 the cocoon, and Fig. 69 the miller.

It is what you might call a sporting proposition and great fun to collect millers. The surest way to get good specimens is to raise the caterpillars from the eggs and feed them upon the leaves they delight to eat. This you will find exceedingly interesting. Another good way is to go out and hunt the big caterpillars; trail them as elephants in Africa were trailed, by the spoor. This develops observation and the same sort of wood-

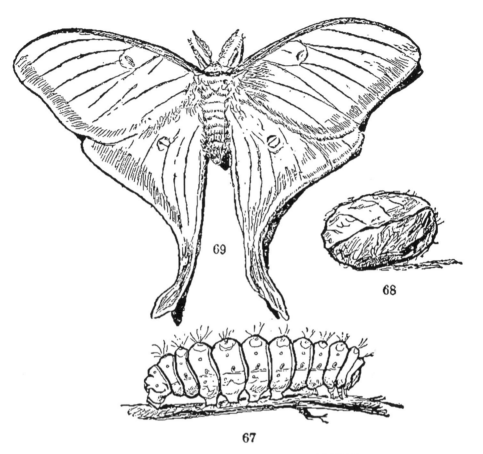

69

68

67

LUNA MOTH, CATERPILLAR, AND COCOON.

craft for the hunter as that possessed by the Indian. The big caterpillars of the giant millers, as a rule, feed upon shrubs or trees, and their droppings may be discovered beneath their pasture.

LUNA MOTHS OR MOON MILLERS

The handsomest of all our millers is the Luna or moon miller, the big, pale-green, swallow-tailed miller which comes from a great juicy caterpillar, the sort of caterpillar that makes a woman " throw a fit." Of course by this we do not mean that the ladies will fall down on their back, kick their heels, and froth at the mouth whenever they see a baby moon miller, but many of the ladies do squeal, and make a great fuss at the sight of one of these caterpillars.

I have captured Luna moths in the scrub pines and sand wastes of Georgia out of sight of any oak, walnut, hickory, or chestnut wood; I have caught them on the shores of Lake Erie in northern Ohio, also in New York City. There are plenty of them around my farm near Danbury, Conn., and I have seen hundreds of them in the woods surrounding my log cabin in Pike County, Pennsylvania.

Although these night-butterflies are very conspicuous and measure about five-and-a-half inches across their expanded wings, with each of their posterior or back wings lengthened out to form a tail an inch-and-a-half or more long; still there are many people who never saw one. That is not all; they never will see one unless some one of you boys shows them a specimen.

The larva, caterpillar, or baby of the Luna miller (Fig. 67) eats from the time it hatches from the egg until it grows to a great fat caterpillar the size of your index finger. It then turns pink, or flesh-color, and gets ready to spin its cocoon. Then it stops eating forever!

No, it does not die, it simply stops eating. Of course, when the pupa, or chrysalis, is locked up in a cocoon, it cannot eat. When, later on, it cracks the pupa shell and crawls out a winged insect, it is too dainty and too beautiful to engage in any such common and vulgar pastime as eating. It simply lives on what it ate while it was a common despised worm.

When the mother moon miller lays her eggs on the underside of leaves or on twigs, the eggs are as white as those of white Leghorn hens, but later on they turn gray. When very young, the caterpillars are a sort of yellowish green, with the last division of their bodies, called the anal plate, of a bluish tinge. The caterpillars feed on the leaves of the oak, hickory and walnut, and I have seen them on chestnut trees. They become fully grown by the end of July.

The larva is of a pale and very clear bluish-green color. It has a yellow stripe on each side of the body. It is addicted to warts, and there are as many as six pearl-like warts of a purple or rose color on each ring of its sausage-shaped body; like the warts you often see on a person's face, the ones on the caterpillar are furnished with a few little hairs. When the caterpillar is not stretching itself, it is nearly as thick as your thumb; it is then a short, stumpy creature, but when walking it will stretch to three or more inches in length.

When kept in confinement, these caterpillars are subject to a sort of spotted fever. Oftimes black spots will appear on their bodies, and then they will die. But if they live to the age of fifty-five (days, not years) they will turn pinkish or flesh-colored.

At this stage of their growth I have often seen them in the beaks of scarlet tanagers (Fig. 70), which fact tells us that the scarlet tanager is one of the agents whose duty it is to keep these caterpillars in subjection. The usually brilliant red bird at this season of the year is moulting, but that does not interfere with its appetite. The bird's plumage has a moth-eaten appearance and some may think that its appetite is as disreputable

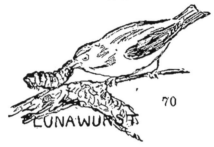

70

as its plumage. I have seen tanagers seize the great fleshy caterpillars of the Luna miller, pinch and squeeze them with their bills, maul them on the limb of a tree, until the whole inside was reduced to a jelly-like liquid. Then the bird would insert its bill into the body of the larva and drink the contents with the same symptoms of delight that a boy shows when sucking an orange.

The Luna is a beautiful, graceful, and artistic moth. The scarlet tanager is one of the handsomest, if not the most beautiful, of our northern birds.   Nevertheless there is no accounting for tastes, and we must own that the baby Luna miller, lunawurst, does not look appetizing to every one.   It is even doubtful if the most enthusiastic of naturalists would be tempted to eat one, although many will eat leberwurst, which looks no better (Fig. 71).

When the Luna miller spins its cocoon, it either draws a few leaves together on a tree, then makes

LEBERWURST
71

its thin cocoon between the leaves, or it creeps down the trunk of the tree and wanders off among the leaves on the ground, and there spins its cocoon. At any rate, after the leaves have fallen in the autumn, the cocoons may be often found by raking up the leaves under the tree.

There is something queer about the Luna's cocoon.   It is noisy!   It sounds funny to say that a cocoon is noisy, but apparently the pupa or chrysalis inside the cocoon is nervous or impatient, and kicks against the confinement in its cell, wiggles and squirms in its prison, so that a lot of cocoons stored away in a box will sometimes produce a noise like that made by shrews, star-nosed moles, or white-footed mice, when they are searching among the dried leaves for food.

Like all the millers or moths, the Luna has many enemies; but I was surprised to find that the dragon-fly or devil's darning needle was one of them. A few years ago I saw a big devil's darning needle make a dash and capture a big Luna miller while the latter was in flight.

The cocoon of the Luna has not the loose end possessed by the cocoons of some of the other big millers. The Luna is sealed inside its cell; but it possesses a special chemical fluid which it uses for softening the threads of which its prison is made, so that it can work its way through the soft spot.

The family to which the Luna belongs, as a rule, spreads its wings when at rest. Very few of them fold or turn their fore-wings backward so as to cover their hind-wings and their bodies. None possess the hook-and-eye arrangement for holding the fore and hind wings together which you find among some of the other moths.

There are several kinds of caterpillars in America which spin excellent silk and it has been claimed that the silk of the American silk-worms is every bit as good as that of the celebrated Tussah silk-worm of India and China, the Pernyi silk-worm from Manchuria or the Yama-Mai of Japan. The Luna moth, or caterpillar, belongs in the family of American silk-worms, but the Luna caterpillar is stingy with its silk and makes a thin cocoon.

All the caterpillars of the American silk-worm family are as naked as a September Morn, but not nearly as pretty, because they have as many warts as an old witch and these warts have short hairs or branching prickles on them. Some of the caterpillars make their cocoons on the ground and some of them fasten their cocoons to the branches of the trees, as does the

GIANT CECROPIA MILLER

There is another giant miller of the silk-worm family and one that is more generally known than the beautiful Luna miller, and this is the Cecropia, whose big brown cocoon we see lashed to the bare twigs of the maple and other shade trees in the winter time.

The Cecropia moth (Fig. 74) is larger than the Luna; some specimens will measure six and one-half inches across the wings. The hind wings are rounded and do not end in tails like those of the Luna moth; the general color is of a reddish brown; they are very fuzzy and their bodies look as shaggy as those of Shetland ponies.

In the middle of each wing is the peacock-feather eye. You will find this beauty spot all through nature; the Luna moth has it, but not so well marked as the Cecropia. Some fishes have it, also some flowers and birds. The jaguar and the leopard have it on their fur; in fact, it is used so frequently that one is almost tempted to think that it is Old Mother Nature's private seal, totem or brand.

The Cecropia caterpillar (Fig. 72) is another green sausage-shaped creature generally classed by the boys as a " tomato worm " or " tobacco worm," that is because all caterpillars look alike to the boys until they have made a study of them.

The baby Cecropia feeds on box elder, apple,

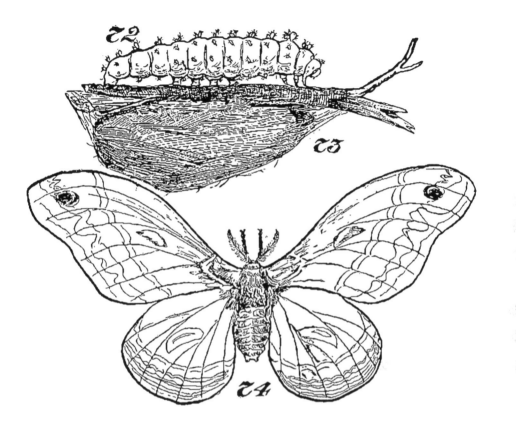

Cecropia Caterpillar, Cocoon and Miller

wild cherry, maple, willow and plum trees and currant and barberry bushes. The mother Cecropia will lay between three and four hundred eggs in a week's time. She lays her eggs on leaves, those watched in confinement being said to have deposited their eggs upon the upper side of the leaves.

The eggs are of a pinkish-white in color, more or less daubed with the reddish-brown glue with which the mother moth sticks the eggs to the leaves. It takes a little over two weeks for the eggs to hatch.

The young caterpillar is of a decidedly yellow color and has a row of warts on its back looking like minute specimens of the Southwestern cactus plants set in a garden row. The baby is full grown by the first of September and will then measure three or more inches in length; it is entirely of a light-green color and it has two balloon-like red warts, studded with a dozen short black bristles, located on the second ring; the two warts on the top of the third ring are a little larger but otherwise the same as the ones described. Then come the yellow warts, egg-shaped and bristled. On the eleventh ring there is one big wart and on each side of the body there are two rows of light-blue warts.

Warts seem to be the favorite decorations with these creatures. There is another row of warts below the blue ones on the first five rings.

In the insect world the female is the bigger and more important of the two sexes, as she has to furnish storage room in her body for hundreds of eggs and see that at last the eggs are properly placed. Like the Luna miller, neither Mr. nor Mrs. Cecropia eat any food after they become moths or millers. The female is so much heavier and stockier than the male that you can ofttimes tell by the weight of the cocoon whether it is going to produce a female or male moth. The cocoon (Fig. 73) is a well-finished silken sleeping-bag, the outside is waterproof and inside of that is a loose fuzzy silk to keep out the cold and then the rounded egg-shaped cell which contains the mummy-like pupa or chrysalis.

These cocoons must be well made because the caterpillars go into them, change to the pupa form and stay there through all the storms of winter; the moth does not come out until the next summer. But even then she would not be able to get out of her prison cell if Mother Nature had not provided her with some of that chemical fluid which softens the glue and threads of the inner cell so that the insect can push its way through.   The small end of the outer cocoon is very loosely woven and the threads hold together, or rather spring together, with their own elasticity, so that all the moth has to do is push her head against them from the inside and crawl out into the big, big world!

POLYPHEMUS MILLER

The Polyphemus miller (Fig. 75) is of a shade between yellow and brown; it also has Mother Nature's beauty spots, or peacock-feather eye spots very distinctly marked on its hind-wings; they are transparent and called window spots; there are also smaller ones on its fore-wings. The band around the front margin of the fore-wings and near the outside edge of all the wings is of a grayish color. Near the outside edge of both pairs of wings is a pink-edged dusky band. There is also a disjointed reddish line with white or pink edges running across the fore-wings. The transparent window spot has two panes of glass in it, so to speak; that is, it is divided by a vein running through it and is enclosed in a window sash composed of yellow and black rings.   The wings themselves spread between five and six inches.

There is no race suicide among the Polyphemus moths.   In Vol. 1 of the American Naturalist a writer tells his experience in raising a million of

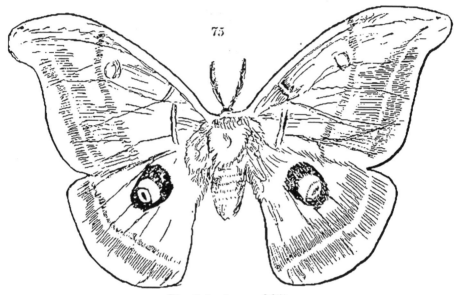

75

The Polyphemus Miller.

these caterpillars in one season!   Mamma Polyphemus either lays her eggs one at a time, or two or three together, usually sticking them to the under side of leaves.   The eggs are larger than the Cecropia eggs and it takes about two weeks for them to hatch.   When the baby Polyphemus hatches out of the little egg, it sometimes runs around like a newly hatched chicken, with **part** of the egg-shell still on it and it often has to turn around and grasp the fragment of the shell with its teeth (?) before it can pull it off.

The babies have five suits of clothes before they go into their cocoons, that is, they change their skins or molt five times and each time they get a new suit of clothes it is several sizes larger than the old suit.

You see it is this way: These little children eat greedily and grow rapidly, but the skin does not grow and at length it becomes too tight to hold them; then they crack it open and crawl out; a simple thing to do, but "All the king's horses and all the king's men could not put them back" into their skins again.

The little babies at first are inclined to be reddish in color, but after changing their clothes they assume a greenish hue, bluish-green above and yellowish-green below. The Polyphemus caterpillar is another one of the " German sausage " type (Fig. 76). It is green and has bias or diagonal white stripes on its sides. On the last division it has purplish-colored V-shaped decorations. All the different divisions of the body are decorated with yellow warts, which naturalists call tubercles. As the babies grow, their colors vary, but green is the constant tendency.

When you disturb one of the big caterpillars it will gnash its teeth like a woodchuck or a bull

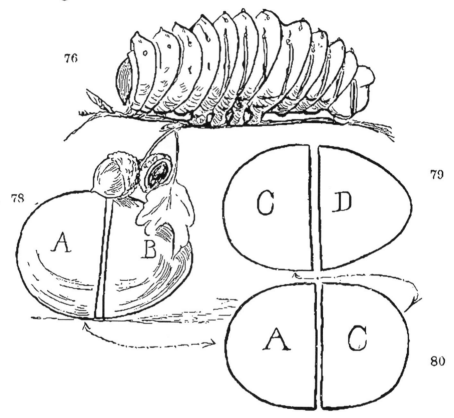

elk. Of course you know that the caterpillar has no real teeth, but the horny parts of its mouth with which it bites, known as mandibles, are rubbed together, making a grating noise like that produced by gnashing teeth.

If the writer has hinted that some caterpillars

are not very handsome creatures, or has suggested that the ladies are afraid of them, he must take it all back because his attention has been called to the fact that two splendid women, keen observing naturalists, Ida Mitchell Eliot and Caroline Gray Soule, both declare that the Polyphemus baby is *very pretty* indeed, and that caterpillars w i t h lustrous red warts are especially clean looking and attractive. That is fine! Attractive caterpillars! Well, this shows that it is unfair to lump the ladies all in one bunch. The writer humbly apologizes for making fun of either the women or the cater-

77

Polyphemus cocoons.

pillars and owns up that both are beautiful.

The Polyphemus caterpillar feeds on the leaves of the plum, elm, apple, maple, basswood, butternut and oak trees. The cocoon (Fig. 77) is made of one silken thread and it is not difficult to unwind it. The cell is " oval cylindrical " and covered with a kind of white powder. The cocoon is usually attached to a curled leaf or two. This is done either on the ground or the trees from which the leaves fall. If the reader does not know what "oval cylindrical" means, he can find out for himself by cutting two hard-boiled eggs (Figs. 78 and 79), in half and then taking the halves with the blunt or biggest ends and fitting them together (Fig. 80). This will make an egg with both ends alike—an egg which might be called "oval cylindrical."

PROMETHIA MILLER

When you are on a hike in the winter time, fall, or early spring you can find the cocoons of the Promethia moth hanging to boughs and branches (Fig. 83), to which they are attached by stems of pliable silk. These cocoons are easily plucked by breaking off the twig to which they are attached and are a favorite specimen with young collectors, who take the cocoons home with them, put them in a vase or some receptacle on the mantelpiece and leave them there until beautiful moths come out.

The millers (Fig. 81) are dark, blackish color with very faint transverse lines and a spot near the centre of each wing, sometimes very faintly marked

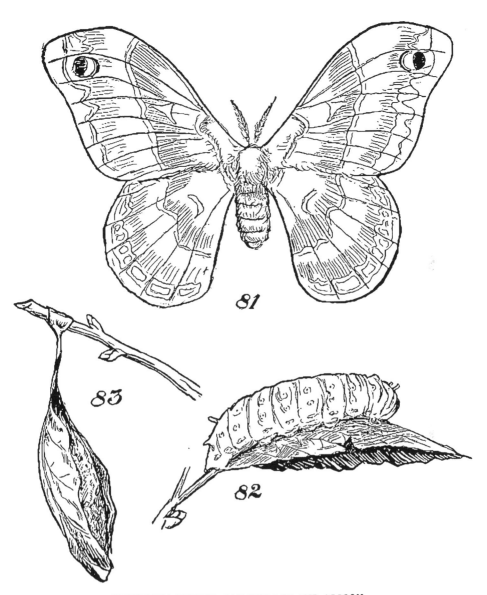

PROMETHIA MILLER, CATERPILLAR AND COCOON.

and sometimes not at all. The front wings of the male differ somewhat from the front wings of the female, the apex, or point, having more of a hook to it.

The caterpillars (Fig. 82) grow to be about two inches long, are of a pale bluish-green color with the legs and shield of yellow; they have shiny black warts, except on the second and third front divisions, where there are coral red ones. There is a wart of similar size and yellow in color on the eighth amidship division, or abdominal segment. The caterpillars feed on the lilac, ash, wild cherry, azalea, and button-bush. The eggs are pinkish-white and are deposited in single rows.

CYNTHIA MILLER

A rose by any other name would smell as sweet and so would a Cynthia caterpillar. Cynthia, however, is a pretty name in itself and the baby Cynthia larvæ are pretty babies (Fig. 84). They have a white bloom on their bodies like a ripe plum and give out a pleasant odor. But the bloom and the fragrance would still be there, even if we called the things " worms." We must not do that, however, because the caterpillar is not a worm; one might

'ust as well call snakes eels because both have long
wiggly bodies or call earth-worms caterpillars.
You see most people talk too carelessly to describe
a thing accurately.

The Cynthia lays about three hundred and fifty
white eggs early in May and in common with other
mother millers sh sticks her eggs, like postage
stamps, to the leaves or branches, using a brown,
gummy glue of her own for the purpose and carelessly smearing her white eggs with it. In two or
three weeks' time the baby Cynthias hatch out and
begin to eat and change their clothes and eat more
and change their clothes more often as they grow.

Not only are the Cynthias handsome caterpillars, but they are also so economical that they do
not like to waste anything, so in spite of the fact
that all the leaves of the trees are handy for them
to eat, they always eat up their old suit of clothes
rather than throw it away. Like Robin Hood, they
dress in green and their costume is ornamented
with black dots, a white bloom and a row of white
tubercles (Fig. 84).

The Cynthias have six legs up in the bow, so
to speak, then a bunch of soft, fat piano-stool sup-
ports amidships and a pair of soft props at the

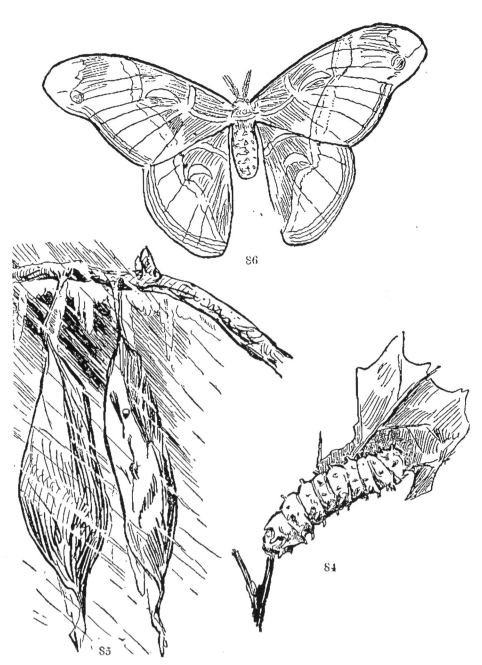

THE CYNTHIA CATERPILLAR, COCOON AND MILLER.
84, BABY; 85, WATERPROOF COCOONS; 86, PERFECT INSECT.

stern. The piano-stools and the stern props are not considered to be real legs, and they disappear when the caterpillar turns into a moth.

The Cynthia cocoons (Fig. 85) are bound to the twigs by yellowish-white silk ribbons, the twig to which the cocoon is attached being itself first carefully wrapped for many inches with silk, then the leaves and leaf stalk holding the cocoon securely bound to the twig. Great bunches of these cocoons often hang together; sometimes there will be a cluster of as many as twenty cocoons on one small branch. In these swinging sleeping-bags the pupæ spend the winter safely protected from the storms of ice, sleet and snow, but not from all foes, because the hairy woodpecker may sometimes be seen hanging on to a twig hammering away on the silken covering of the cocoon. The sharp beak of the woodpecker makes a hole through the cocoon's walls and the skin of the pupa itself, then the bird laps or sucks out all the insides and leaves only the dry shell.

The Cynthia moth (Fig. 86) could not join the Sons of the Revolution nor the Colonial Dames because he or she does not come of early American stock. The Cynthia originally was a Chinaman, but like all other immigrants he has made himself at home here, and although I believe the Cynthia fed exclusively on the ailantus tree in China, it will feed here on the sycamore, spicewood, dogwood, plum, wild cherry and other leaves. As a boy, the writer never called these moths Cynthias, he only knew them as the ailantus moths, but Cynthia is a good name for them and one easily remembered.

The miller measures from four and one-half to nearly six inches from tip to tip, is a sort of olive-green in color, peppered over with black scales, with a lilac band across the wings and the other bands white with a tinge of lilac. The half moons or crescents are yellow and nearly transparent. They have nature's beauty spots, the peacock-feather eye spot, near the tip of the fore-wings and the body has white tufts on it.

The millers do not eat; they could not if they tried to do so, because their mouth parts are un-finished. They have no tongue, or the tongue they have is a sort of a make-believe affair, a hold-over from the time when they once had tongues which now are of no more use than the buttons on the back of a man's coat.

IO MILLER

"Once on a time, when dogs ate lime, and peacocks chewed tobacco," there lived a certain goddess whom the Romans called Juno, the Greeks

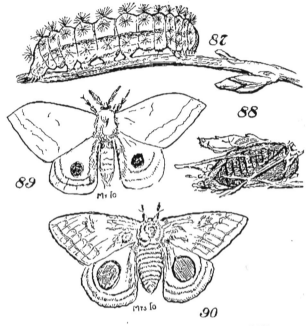

Mother, Father, Baby and Cocoon of Io Moth Miller.

called Hera, the Etruscans called Uni and some other pagans called Io. This same Io or Hera was of strong, hearty, rebellious character and full of intense hatreds; that is, according to the poets, and the poets are about as truthful as the ancient religions handed down by the priests of those days. None of their stories would hold in court to-day. Some say Io was the sister of Zeus, some say she was the wife of Zeus and others that she was the wife of the Egyptian king named Osiris. But of one thing we are certain, Io is classical and hence deserves to have something named after her, and it is probably because Io was a dangerous woman that the dangerous caterpillar is called Io.

When we say "dangerous caterpillar," the reader must not think that the caterpillar is going to bite or kill the collector, but it can make it very disagreeable for any one who handles it because of the poisonous quality of the prickles on its back (Fig. 87).

The baby io is hatched from peg-top-shaped eggs, which are deposited in groups on dog-wood, sassafras and bayberry bushes. The caterpillars do not look like those previously described, the big harmless caterpillars, and the io does not suggest a sausage, but scientists have placed it in company with the big silk-worms.

The caterpillars are pea-green in color and decorated with hedgerows of green poisonous prickles. They have dark brown stripes, edged below with white on each side of their body. The stripes begin at the fourth department, segment, or ring, and end at the tail; the green thorn bushes on its body are tipped with black and are all of the same length; there are about thirty thorns to each bush, all springing from a common centre, and there are about six of these bunches of stinging spikes on most of the rings. But on the last two rings there are only five and on the first four there is an additional cluster on each side near the bottom.

The pretty io moth, butterfly or miller (Figs. 89 and 90) is much smaller than the giant moths already described, but for other reasons it is placed in the group containing the giant silk-worms. When I was a big boy, by much research and inquiry I found that the name of this miller was then Saturnia io, but there is not much use in loading one's head up with scientific names because they do not last long; since then I have heard it called Hyperchiria io and Antomeris io, and by the time my readers are as old as the writer it may be called after Jupiter's spectacles or Mars's binoculars, but io will probably stick, so we will

7

call it the io miller, which is the grown-up io caterpillar.

Mr. Io (Fig. 89) is smaller than his wife (Fig. 90) and not so gaudy, but he has nature's beauty spots on his two hind wings, these spots, however, being much brighter and larger on Mrs. Io. The gentleman and lady differ both in color and size. The gentleman is of a deep Indian yellow with two wavy lines running bias across its fore-wings toward the back edge, zig-zagging near the bottom; these lines are of a reddish-purple color.

The back wings or second wings next to the body are purplish-red, and near the back edge there is a curved band of the same color. The beauty spot is made by a big blue blot with a black border and a simple dash of white, and the beauty spots on the skirts of lady io are much larger than those on gentleman io's coat-tails. When these moths are at rest they fold up their wings over their back, making a roof like that of a house, in place of spreading them out flat, as do the moths previously described.

The caterpillar spins a cocoon (Fig. 88) on the ground, picking up leaves and rubbish of any kind and fastening it to the cocoon. The cocoon is thin, of a very gummy brown silk, and as soon as the cocoon is finished the caterpillar is changed to a chrysalis and thus it sleeps all winter, coming out about June or July the next summer to mate, lay eggs and hatch out a new crop of poison caterpillars, tiny little fellows with black heads and tan-colored bodies who march along like soldiers following a leader to get their rations and march back again when they are through. They keep up this military formation until they think they are big enough to go off scouting on their own hook.

CHAPTER FIVE

AMERICAN ROYALTY

ALL you boys who read American history know that on December 16, 1773, the British government took the taxes off everything except tea, then tried to force the Americans to become tea drinkers. The result is that the United States has ever since been a coffee-drinking country!

A lot of our ancestors dressed up as Indians and threw all the tea overboard in Boston Harbor. The British Admiral Montague poked his head out of the window, as the make-believe Indians went by, and cried, " You've had a fine night for your Indian caper, but you will have to pay the fiddler yet!" To which our husky forefathers replied, " Just come out here, and we'll settle the bill in two minutes."

You boys should be proud of those ancestors; they were a spunky lot and when they threw the tea overboard they thought they threw everything relating to kings and royalty with it, but they were mistaken, for right here in America we have native-born emperors and a royal family!

With a republican form of government and under the democracy of Thomas Jefferson, this royal family thrives and no one begrudges them their title and no anarchists throw bombs at them. They are forest kings and belong to the royal family of millers; they are first cousins of the giant silk-worms.

The caterpillars have horns on the second and sometimes on the third divisions of their bodies. They live on the leaves of the forest trees and bury themselves in the ground when they feel the change coming over them warning them that they are soon to take the chrysalis form.

EMPEROR MILLER

The first member of this family is the Emperor miller (Fig. 91). It has a spread of wings of four to five and a half inches, and is a beautiful sulphur yellow with purplish or violet color specklings or markings. I believe its present scientific name is Eacles Imperialis, but you need not try to remember this name, for, although it is its scientific name to-day, no one can tell what its name will be to-morrow. But the common name, Emperor moth, or Emperor miller, will probably stick to it always.

The baby Emperors are good-sized caterpillars, which are ripe, so to speak, in autumn (Fig. 92). Sometimes the caterpillars are brown and sometimes green, sometimes hairy, but more frequently look like the one shown in the illustration. The chrysalides or pupæ are black, stockily built and armed with the spines or prickles which help them to wiggle up to the top of the soil when the miller wants to get out of its mummy case. The baby millers will eat the needles of pine and hemlock, also the leaves of oak and birch, sweet gum and sassafras, hickory and numerous other wild leaves. The eggs are laid in June after the moth comes out of the chrysalis and mates; she sometimes lays the eggs before she mates, but of course these eggs do not hatch. The eggs are large and yellow and stuck singly on the upper side of the leaves. They hatch out in about two weeks' time.

The Emperor miller is beautiful and one that you should, by all means, have in your collection. As it is not very rare, it would be well to mount at least three of them, an Emperor and an Empress, with their wings extended and then another, either an Emperor or an Empress, with its wings half folded in their natural position while at rest.

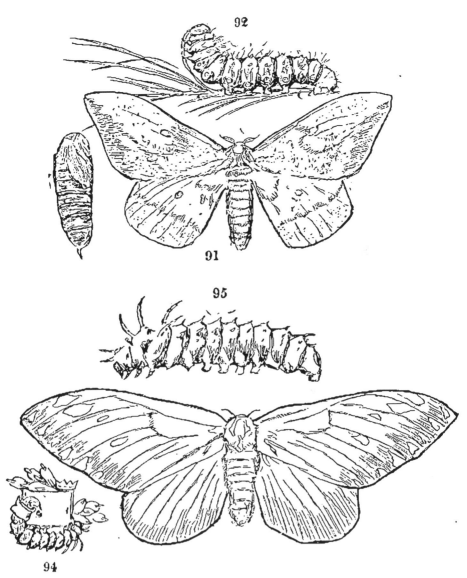

EMPEROR AND REGAL MILLERS WITH THEIR LARVÆ.

REGAL MILLER

The next member of this royal family is the Regal miller. These millers deserve their royal title, as they do nothing for a living except eat and wear good clothes, in truth, they wear beautiful clothes. The Regal moth's name is Cith-e-ro′ni-a re-ga′lis to-day, to-morrow is may be Kaiser or Tzar regalis, but it will probably remain the Regal moth to-day and to-morrow.

A scientific name is easily cut out and forgotten, but the common name is difficult to change; it becomes part of the folk-lore and is not easily forgotten and I will wager that not one of my readers will forget the name of the Emperor miller nor the Regal miller after he has caught one of these beautiful specimens, identified it and placed it in his collection.

Citheronia, a Greek poet, and Regalis, royal. Thus you see this might be written, a royal Greek poet, but, if the royal Greek poets had horns like the caterpillar of the Regal moth, they must have been more comical than handsome.

The Regal moth lays large amber-colored eggs, very much like those of the Emperor moth, although they are somewhat bigger and bordered with a red line. They hatch out in between two and three weeks' time. The babies are fond of butternut and ash-tree leaves. When they take off their old suit of clothes, they eat it.

The caterpillar will grow to five inches in length; it does not hold on to the twig with its props (tail props) but hugs the twig with them. Some time in August they go into a pupa or chrysalis form.

Although the Regal miller (Fig. 94) does not appear to be common anywhere, it always attracts the attention of any person who meets it, and hence it has local names. It is sometimes known as the walnut miller, and the caterpillar is often called the "horned hickory devil" (Fig. 95), but the horns are only a bluff, they do not sting or hurt you. The front wings of the moths have yellow spots on a sort of an olive-colored background with stripes of lead color between the veins of the wings. The body of the insect is a yellowish brown with yellow markings, the feelers or antennæ a bright orange color with a tinge of brown. The moth will measure about six inches across the wings.

There are a number of princes and grand-dukes and all those sorts of things belonging to this family known as the oak caterpillars. These millers are much smaller than the Regal or Emperor. The dotted miller is an example. There is some suspicion of there being a stinging quality to the horns of the dotted miller's caterpillar. These millers are not beautiful. They are brownish in color and you will know them by the white dot on the forewings.

CHAPTER SIX

SPHINX AND HAWK MILLERS, JUG-HANDLES AND TOBACCO
WORMS. NOTCH-WINGED MOTHS

SOME of the caterpillar family have acquired
the drug habit, and are what the newspapers would
call " dope fiends," but the poison seems to agree
with them and does not affect their health or their
nerves; they wax fat upon a diet of tomato leaves,
tobacco leaves and potato leaves, all of which we
know are exceedingly unwholesome and dangerous
for human beings to eat. These caterpillars, however, even devour the poison leaves of the jimson
weed (Datura).

Jimson weed is not well known up North and
it is only of late years that it has appeared around
New York, but it is the common weed of the vacant lots in the Ohio valley. It has a prickly pod
of poison seed and a morning-glory-shaped blossom. The blossom is a great resort for bees, which
may easily be caught when they enter the flower,
by pinching up the end of the flower and imprison-
ing the insect.

The boys call the caterpillars of the Five-

Spotted Sphinx tobacco worms, potato worms and tomato worms. During slavery times the negro boys picked these caterpillars from the tobacco plant and the overseer, following them, made the black boys *bite the heads off* of all the caterpillars that they had passed unnoticed. This, my colored

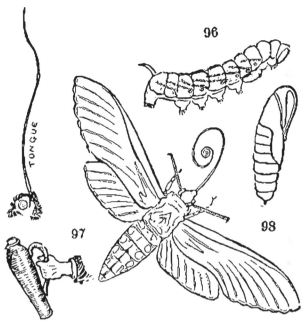

The Jug Handles.

informant told me, made the boys " powerful careful."

The caterpillars have diagonal or bias stripes on the sides and a horn on their tail called the caudal horn, but the part about these creatures which interests the boys most is the fact that the pupæ of many have a curious jug-handle to them. This jug-handle (Fig. 97) is really the case which holds and protects the long tongue of the moths, or the hum-bugs, as the children call them. Among the naturalists they are known as the hawk moths or Sphinx moths (Fig. 98).

The reason they call them the Sphinx moths is because the caterpillar rears up its head so that it reminded Carolus Linnæus, the naturalist, of the big stone sphinx head sticking out of the sand in Egypt. The caterpillar is a large green crawler, which grows as thick as one's finger and three inches or more in length and reaches its full growth between the middle of August and the first of September (Fig. 96); then it crawls down the plant and buries itself in the ground, where it changes to the "little brown jug" form shown in Fig. 97.

The funny part about these changes which all caterpillars are in the habit of making is that they all occur inside the skin, then the outside skin breaks and a new creature, entirely unlike the old one, wriggles out of the crack, just as the butterfly comes out of the skin of the chrysalis.

There are a number of moths belonging to the jug-handle family but some have short handles

pressed up against the body of the chrysalis, as is
the case with the Pen-mark miller's pupa shown in
Fig. 99. Figs. 100 and 101 show the caterpillar
and moth of this miller. Some of the relatives have
no jug-handles at all.

The jug-handles are among the largest and
stoutest of the Lepidoptera.
They are the millers which we
see flying around the flowers at
night, and when their tongues are
out searching for honey in the
flowers they look so much like
humming birds and act so much
like these little befeathered mites
that they are often mistaken for
them.

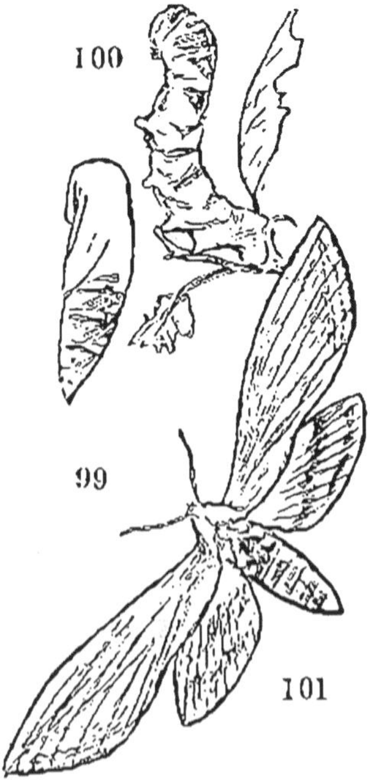

There are between three and
four hundred of these millers.
One kind lives on the pine trees and others on all
sorts of leaves. A great many of them are down in
Mexico, Central America and South America.
June and July is the time you will find them at
home, flying around the flowers in the evening.

Our potato " worm " moth belongs to the
largest ones of this family, and, although its tongue

is longer than that of a village gossip, it is short compared to the nine-and-one-quarter-inch tongue of one of the Madagascar moths. But we will not have room to describe more of these millers. One could make a big book on hawk moths alone.

NOTCHED-WINGED MOTHS

In making your collection of hawk moths, do not forget those with the notched wings. The caterpillars of this kind may be found on wistaria, raspberry, oak, apple, white birch, willow, cherry, hazel and other trees and shrubs and there are quite a number of them. The notched wings may be found flying around your lamp in the farmhouse or hiding under the projections on the outside of the house in the daytime.

The blind-eyed miller (Fig. 102) is a notched-wing which lays bright-green shiny eggs, but the vivid color gradually fades out before they hatch. The babies creep out of the egg shell sometimes in less than a week and sometimes a few days over a week after the eggs are laid.

The little caterpillars when they come out of the eggs are very lively when touched and will stand up on their hind legs and jerk their heads in a threatening manner, as if they were going to do all sorts of things to you, but it is all a bluff. The caterpillars (Fig. 102½) vary from a blue-green to a yellow-green in color and grow to between two and three inches in length. The chrysalis case has no visible jug-handle or tongue-case to it and the pupa is usually nearly two inches

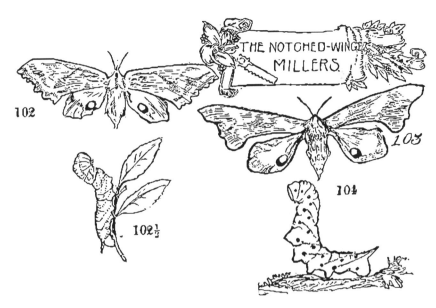

THE NOTCHED-WING
MILLERS

102

103

102½

104

long and stockily built. The millers (Figs. 102 and 103) vary from a brown to a fawn color, the hind wings are pink with an edging of brown, sometimes a pink blush all over them, with maybe a fawn-colored edge. These millers are blondes because they have blue eyes, and Irish blondes because they have very black eyelashes; in other

8

words, their eye-spots are colored blue with a black
border to them; their bodies are fawn-colored and
the gentlemen carry widely pectinated (comb-like)
antennæ while the ladies carry simple antennæ.

There are many other notched-wing moths, but
we will leave you to hunt them and will figure only
one more, the pretty and common Purblind Myops
(Fig. 103). Both these moths, the blind one and
the purblind one, have eye-spots on their hind
wings. The same eye-spot we referred to before
as Nature's beauty spot, but maybe Nature is using
sign language like the Indians and the Gypsies,
and this is her Swastika, her good-luck sign. The
caterpillar to the Purblind Myops (Fig. 104) has
spots along its sides like buttons. Of course it has
a horn on its tail and is fond of rearing its head
and arching its neck, so to speak, like a checked-in
horse.

CHAPTER SEVEN

SUNSHINE MOTHS. CLEAR-WING MILLERS. HUMMING-BIRD
MOTHS. THE WHITE DEATH. FRUIT BORERS
AND SQUASH-VINE MILLERS

CLEAR-WING MILLERS

ALL of us who are interested in insects have, some time or other, been deceived by a thing visiting the flowers in the daytime and having the appearance and actions of a humming bird. But after we have been fortunate enough to capture one of these creatures, we discover that it is not a humming bird, but a "hum-bug;" in other words it is a moth and belongs to the Clear-Wing tribe, the family of moths (Figs. 105 and 107) which are noted for their transparent wings. These millers, when they creep out of their chrysalis or pupa or mummy case, look very much like the hawk moths already described or the members of the Sphinx family of the Jug-Handle tribe.

But the clear-wing moths have their own ideas on personal adornment and they are dissatisfied with their wings when they first emerge, so they buzz around until the coating of scales is shaken like dust from their wings, that is, all except such as are tightly fastened on along the edges of the wings shown in the two types of the clear-wing moth (Figs. 105 and 107).

Some of the smaller moths resemble wasps, some of them look like bumble-bees and some like

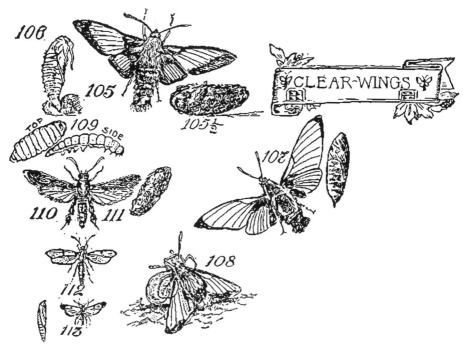

the Ichneumon fly. One of the bumble-bee kind (Fig. 108), is shown in the grasp of a white spider and the drawing is made from a water-color sketch which I painted while watching the white assassin kill the bumble-bee moth. This happened near my log cabin on the shores of Big Tink Pond in Pennsylvania.

I had discovered a milk-white spider concealed in a white flower, where it made a living trap for such insects as the flower might attract. By this means it captured a bumble-bee moth, and the latter died almost without a struggle. A poison secreted by this ghostly spider, known to the mountain boys as the "white death," seems to be stronger than that of the web-making spiders. It may be that as almost instant death is necessary to prevent the victim's escaping when the spider has no web to help him hold the captive a stronger poison is necessary.

In this connection it is interesting to note that a box filled with all sorts of live spiders by a small boy who was making a collection, when left over night was discovered in the morning to have but *one live* specimen in it. The boy found the " white death," or to be more scientific, the female Mi-su-me′na va′ti-a nestling contentedly in the midst of the dead bodies of its victims.

The smaller clear-wing millers are often mistaken for bees, hornets, etc., but as soon as one discovers that they are moths, one knows to what tribe they belong.

Unlike most of the millers, they love the sun-

shine and then they all have funny little tails like
humming birds which they can spread out at will.
The caterpillars are borers—that is, they are the
sort of grubs which eat their way into stems and
roots of plants and feed upon the inside bark, the
wood or pith. Fig. 109 shows the grub or larva of
the squash clear-wing.

The caterpillars to the larger clear-wings (Fig.
105) are very much like those of the Sphinx moth.
Those of the smaller clear-wings (Figs. 109, 110
and 111) make their cocoons of small bits of wood,
and by the aid of their little prickles on their
chrysalis shell they work their way out of the cocoon
(Fig. 106) and also part of the way out of the tree
trunk, if they happen to be in one. When the moth
frees itself from its mummy case, it leaves the latter
sticking half way out of the hole in the wood.

These bee clear-wings or, as Harris calls them,
Ægerians, fly only in the daytime. They love the
bright sunshine and are gaily colored with yellow,
black and red, although some of them are not conspicuous because of the smallness of their size.
Fig. 110 is the squash-vine miller. It has an orange-
colored body spotted with black, a pair of cowboy

chaps on its legs made of long orange- and black-colored hairs. Its wings spread about one and one-half inches; only its hind wings are transparent. Some call this miller the porch vine Ægeria. The cherry-tree miller (Fig. 112) does its most damage when the larva bores into the roots of the trees. The miller has all four wings transparent, but the framework and borders of the wings are steel blue, this being also the general color of the body of the male insect, the wings of which spread about one inch. But his wife, if not the better of the two, is the larger; her front wings are not transparent and she wears a broad, fashionable girdle of orange color around her body and can spread her wings half an inch further than her husband.

The little villain shown by Fig. 113 is an enemy of our pear trees. The wings of this little marauder do not spread much over a half an inch, are fringed and veined with purple-black and the front wings have a wide dark band with a coppery glint to it. The back of the moth is a dusky or purple-black color, its under side is of a golden yellow and it wears a golden collar and golden epaulets. It also has a yellow tail and a yellow girdle across the middle of its body with two yellow rings of the same color.

There are other millers belonging to this family, millers which love to destroy the wild currants and lilac and other plants, useful and ornamental, but we have given these little pests as much room as we can spare.

CHAPTER EIGHT

UNDER-WING MILLERS. TIGER AND LEOPARD MILLERS.
YELLOW BEARS. HOBO CATERPILLARS

UNDER-WING MILLERS

OFTEN when one is walking through the woods
on the look-out for specimens, one may discover
some moths upon the trunks of the trees. When
the wings of these moths are folded, they are in
color and marking so similar to the bark upon which
they rest that they are easily passed by unnoticed.
But the moment they spread their wings, all concealment is lost, for their underskirt, so to speak, is
often a very brilliant and beautiful one. Hence
they are known as under-wing millers.

In making a collection of these millers, one
should secure enough of them when possible to
enable one to preserve some specimens in their
natural position of rest with folded wings on a
piece of the bark of a tree, and others with their
wings extended showing their beautiful underskirt.
Fig. 114 (Catocala relicta) is the gray-backed
under-wing. When the wings are folded it has
the appearance of a piece of gray bark, but when

they are open it shows the under-wing of a sort of chestnut color marked by two white bands. Fig. 115 (Catocala concumbens) is the light-red under-wing. The upper wings or front wings are of a brownish tinge, but the under-wings are red with an outside margin of yellow, then crossed by two dusky bands. Fig. 116 (Catocala ultronia) is the deep-red under-wing, the upper wings of which are darker than those of Fig. 115, in fact the whole moth is darker; it also has a yellow scalloped border upon the edge of the under-wing, the deep-red surface of which is marked by two dark bands. Fig. 117 (Catocala gracilis) is the one-banded yellow under-wing and Fig. 118 (Catocala amica) is the two-banded yellow under-wing. These two moths are very much the same color, but as you may see, they are marked differently. They are all called Catocala moths, unless the name has been changed since the writer collected them. They are, however, still known as the under-wings.

BEAUTIFUL BELLA MILLERS, TIGER AND LEOPARD MOTHS

Around the bed-room lamp in the old farm-house is one of the best hunting grounds for the

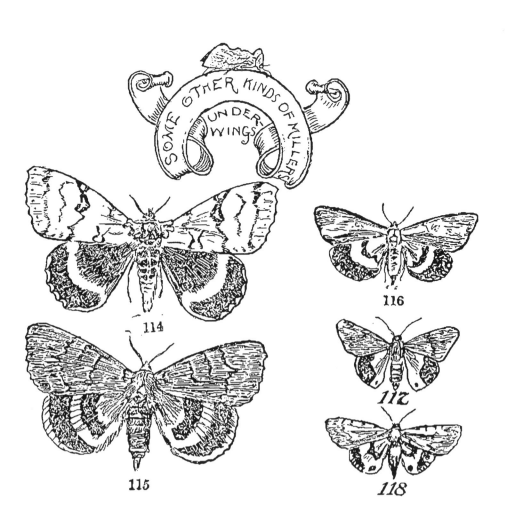

SOME OTHER KINDS OF MILLERS UNDER-WINGS

114

115

116

117

118

collector of moths or millers. These insects seem to be possessed of the idea that they must commit suicide and they will even drop down the chimney of the kerosene lamp.

Among the tent caterpillar moths, under-wing moths, gypsy moths and brown-tailed moths, one will find the beautiful Bella (Fig. 119), a miller that spreads between one and one and three-quarter inches. This so-called tiger moth, with the rest of the group, differs in appearance from the under-wing moth principally because its upper skirts as well as its underskirts are beautifully decorated.

You will find the beautiful Bella any time from the middle of July to the first part of September. It has naked feelers or antennæ. Its front wings are of a deep yellow, decorated with about six white bands and on each band is a row of black dots. Its under wings are light red with a border of black. It has a white body and the thorax is dotted with black.

The caterpillar may be found late in July and August in the seed pods of the rattlebox. It is yellow with black and white rings (Fig. 120) and the pupa or chrysalis remains a week or ten days in that state before the moth hatches. From all accounts the caterpillar seems to favor the Pulse family of plants, that is, plants with seeds in a pod like peas and the beautiful wild blue lupin (Fig. 122).

It gnaws a hole in the pod (Fig. 120), creeps

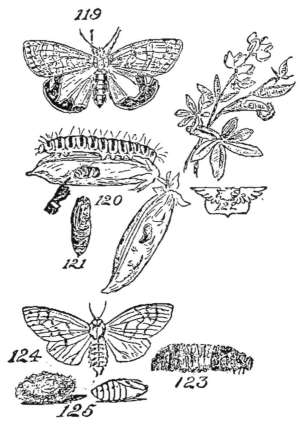

in and hides itself and there, undisturbed by birds or man, devours the green seeds. So well concealed is the caterpillar that the great authority on insects, Dr. Harris, says: " The caterpillar is unknown to me." That means that the Doctor did not find one, but I have no doubt you boys can.

The beautiful Bella belongs with the group of so-called tiger moths, but in reality the spotted ones should be called leopard moths, because tigers are not spotted and leopards are. We will, however, not quarrel with this name because at least one of the moths is called a leopard miller (Fig. 128). They are all of them pretty and add to the beauty of a collection and most of them are easily caught around the lamp at night. The caterpillars, as a rule, belong to the hobo class—that is, they seem to have no permanent abiding place. You meet them hustling along the roadside and in the paths, apparently travelling in any and every direction, and maybe if we could hear them and understand caterpillar language, they would be found to be singing:

> "We-e-l, I ain't got no reg'lar place
> That I kin call my home,
> Ain't got no permanent address
> As through this world I ro-o-am,
> An' Portland, Maine, is just the same
> As sunny Tennessee,
> For any old place I hang my hat,
> Is Home Sweet Home to me."

ISABELLA TIGER MOTH

The Isabella miller (Fig. 124) is a dull yellow with a few black dots on the wings, but every boy knows the caterpillar (Fig. 123). It is the lively crawler, colored black fore and aft, and reddish brown amidships, and is thickly covered with a lot of evenly clipped stiff hair. I discover to my sorrow that in confinement this caterpillar will eat up other more tender caterpillars, although I never knew it to eat caterpillars protected, like itself, with a thick coat of hair.

When cold weather approaches it hides away under boards, sticks and stones, where it remains sleeping until the next spring. In April or May it makes itself a covering, using the hair of its own body to weave into this dark oval-shaped cocoon (Fig. 125). The moths come out in June and July. The wings of the moth expand sometimes as much as two and three-eighths inches. This miller finds a place in this book because every boy knows the caterpillar and is naturally anxious to know what kind of a moth it produces.

THE YELLOW BEAR

The yellow bear, common everywhere in our garden, is a hairy caterpillar. Unlike the Isabella, the hairs are very uneven in length, but because it is so common we must mention it along with the Isabella caterpillar. Almost any sort of vegetable seems to suit the yellow bear's appetite. The moth is a snowy white with seldom more than three dots on each wing.

THE SALT-MARSH MILLER

This is a common white miller with black dots on its wings. Although it is called the salt-marsh miller, it does not confine its attention to meadow lands along the sea-coast. Every boy knows it, but every boy does not know that the male and female millers differ in the color of the wings. The female is a white miller, but the male only partially so. Only the upper part of the fore-wings of the male are white and underneath they are yellow, the hind-wings also being yellow.

THE TIGER-MAID MILLER

The tiger maiden wears a velvet gown of black. The decorations of pink or yellow are formed like

9

cracks in the winter ice, some at right angles to
each other (Fig. 126) and some diagonally running
across the wing.

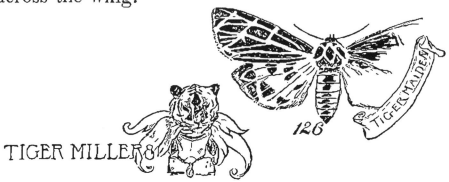

TIGER MILLERS

THE CLYMEME TIGER MILLER

This miller can be easily recognized by the two
dusky spots on its lower wings and the oddly-
shaped dark borders to its upper wings, the wings

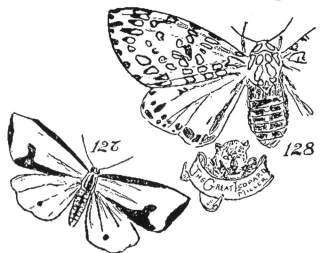

themselves having a body color of tawny yellow
(Fig. 127).

THE GREAT LEOPARD MILLER

This is a beautiful night butterfly (Fig. 128)
of very light color, with brownish red spots on its
thorax and fore-wings; the hind-wings are trimmed
along their outer edge with dark spots and dark
streaks along the outer edge of the lower wings
next to the body. I believe that I have caught all
these tiger moths around the lamp at night, as well
as many others not featured in this book.

CHAPTER NINE

TENT MILLERS

THE tent caterpillar, which forms a large cob-web-like nest on the wild cherry and the haw bushes in latter part of April, through May, in June and July, often spreads from these trees to the orchards, where it is very destructive. I have seen large trees in Connecticut completely denuded of foliage and every branch enveloped in a sheath of cob-web-like silk (Fig. 129). Not only were the branches enveloped, but there were paths running down the trunks of trees out to the grass and underbrush, silken roadways of cobweb material.

The truth is, these caterpillars do not seem to be able to find their way by the stars or the sun, and as they carry no compass they have invented a way of their own for marking the trail. From their mouth they spin out a thread of silk as they creep along; when they want to retrace their steps it is only necessary for them to follow back the

132

line of silk they laid; doing this often makes the well-marked silken trails.

The moths lay their eggs on twigs, surrounding the twig with a cylindrical bunch of from 250 to 400 eggs, placed side by side in perfect rows around the twig and varnished with a gummy mat-

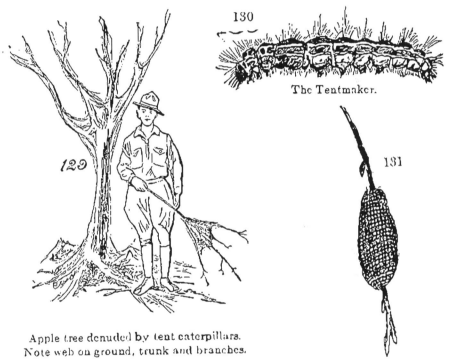

130

The Tentmaker.

131

129

Apple tree denuded by tent caterpillars.
Note web on ground, trunk and branches.

ter which is supplied by the female moths and which waterproofs the eggs (Fig. 131). These bunches of eggs may readily be detected in the winter when the twigs are bare.

As soon as the little caterpillars hatch in the latter part of April and the first of May, they begin to nest in a convenient fork of a tree. The caterpillars all work together to make these tents, which form retreats for them when they are not engaged in eating, and, if you will secure a forked stick and push one of these tents down, you will find it contains a ball of caterpillars as big as your two fists. As the young increase in age and size they enlarge the tent. At certain times, depending upon the weather, they all come out together to eat and, when their feast is finished, they all retire at once.

When fully grown the caterpillars (Fig. 130) measure about two inches. They have black heads and a black back. From one end to the other is a whitish line on each side of which, on a yellow background, are a number of fine crinkled black lines that, lower down, mingle together and form a broad black stripe, or rather a row of long black spots, one to each ring, in the middle of which is a small blue spot. Below this is a narrow wavy line and lower still the sides are variegated with fine intermingled black and yellow lines which are lost at last in a general dusty color on the under side of the body. There is a small dusky wart on the top of the eleventh ring and the whole body is thinly covered with short, soft hair. Some time in June the caterpillars leave their nest and travel restlessly, often creeping on one's clothes and not infrequently entering the house in search of sheltered crevices where they can spin their cocoons (Fig. 134).

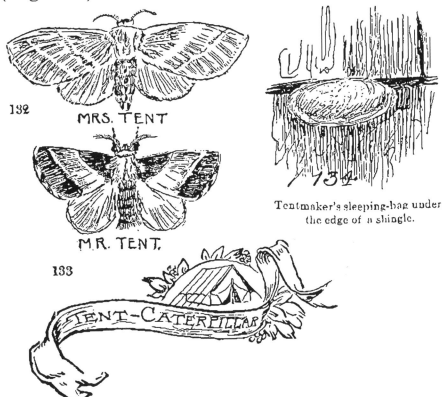

132 MRS. TENT

M.R. TENT,

133

Tentmaker's sleeping-bag under the edge of a shingle.

TENT-CATERPILLAR

The cocoons are made of loosely woven silk plastered over with a thin paste which when dried is like lime, so when one mashes a cocoon the paste turns into a dust of a pale yellow color. Two weeks longer are spent in the chrysalis state before

the moth cracks the mummy case and works its way
out through the wet and softened end of the cocoon,
dries its crumpled wings and assumes the form of
Fig. 132 or 133.

ARMY " WORMS "

Every once in a while some section of the country is invaded by an army of caterpillars known
as ARMY " WORMS," but when people call a caterpillar a worm they are talking loosely. We have
said something about this before, but we refer to
it again because we want the boys to know the
difference between a caterpillar and a worm. The
worm family is such a big one and has so many
distant relatives included in it, that I find it almost
impossible to give you a definition. One scientist
says, " As a rule, worms are bilateral, segmented
animals with the nervous cords either separated
or united by commissures, and resting on the floor
of the body," and so on. But I do not believe this
will help you much to understand it. If, however,
you will catch an earthworm and compare it to a
caterpillar, you will immediately see differences
which are more easily detected than the meaning
of the words just given.

The trouble is, boys, that the English language

is composed of a whole group of languages. There is a printed language, a spoken language, the language of the biologist,, the language of the doctor, the language of the surveyor, and the language of the electrician, etc., but not many of these fellows understand each other's language. Then there is the language of the boys, which very few grown people ever use, and the scientist is one who does not use it. But all this is to tell you that the army " worm " is not a *worm!*

When these caterpillars start a campaign, they take no provisions with them, but live on the country. They will strip every vestige of green from the fruit top of the oats, rye, wheat and timothy (Fig. 138) leaving only a straight, bare stalk standing.

The sketch you have with this (Fig. 135) is one I made from the live caterpillar while it was chewing off the end of a timothy stalk. Fig. 138 shows a head of grain before and after the visit of

the caterpillar. It is very lucky for the people that
the army " worms " do not often visit us.

The moth is an ordinary looking miller (Fig.
137) of a shabby yellow drab or russet color, small
white dots near the centre of the front wings and a
dusky bias stripe around the tips. It is not quite
an inch and three-quarters from tip to tip. The
fore-wings are freckled with black and crossed by a

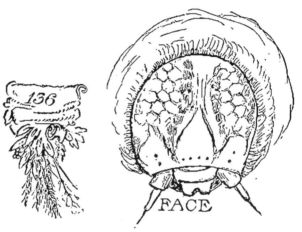

row of black dots a short distance from the hind
edge, one on each vein.

This row of dots when it reaches the middle
of the wing curves forward, making a dusky stripe
to the tip, the wing being slightly paler and yellowish along the side of the streak of dots. The milk-
white dot in the centre of the front wings is placed
upon the mid-vein, but all the markings are indistinct. The hind-wings are a smoky brown with a

purplish-blue to them, the veins almost black and the wings nearly transparent.

The full-grown caterpillar is shown by Fig. 135, while Fig. 136 shows enlarged view of the face of the caterpillar. The army "worm" sometimes measures two inches in length and is about as thick as a quill toothpick. Kill the moths and kill the caterpillars whenever you see them. Preserve specimens so that you will always know them. Note in your record book the names of the different

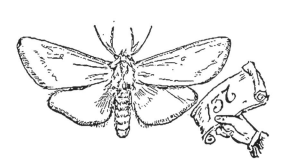

birds that you see feeding on them, and when you say your prayers at night, in place of asking the Lord to give you something which might not do you any good if you had it, thank Him for not sending any more army caterpillars and for supplying us with birds to keep them in check.

CUT-WORM MILLERS

The tent caterpillar is a nuisance, the army worm an aggravation, but the meanest, most unprincipled, disreputable caterpillar among the inhabitants of the orchard and garden is the cutworm! This disagreeable, dark-colored, hairless caterpillar lies hidden in the ground waiting for one to set out a row of tomato plants, young cabbages or anything nice in the vegetable line in which one takes great pride, and then at night he sallies forth and bites off all the stems near the surface of the ground.

Cock robin helps keep these fellows in subjection and eats great numbers of them, but the only safe way to protect the young plants from the cut-worm is to put a little collar of stiff paper around each stem, allowing the lower edge of the collar to extend down into the ground.

I might say more things about cut-worms (Fig. 139 and 141), but they are no friends of mine. I do not like their methods, in fact I do not like their character; the cut-worm is not a fit associate for decent people and I rank it with men who poison pet dogs.

Most of the moths (Figs. 140 and 142) appear

in midsummer, along in about July or August, then
they proceed to lay their eggs in the gardens, in
the meadows and the ploughed fields. Upon the
approach of winter the caterpillars (Fig. 139),
curl themselves up and sleep until the next spring
down in the earth below frost.

As soon as the dirt begins to warm up a bit,

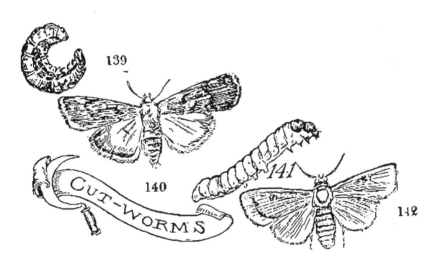

they work away towards the surface and watch for
you or me to set out our potato vines, tomato vines,
nice rows of succulent-stemmed cabbage plants,
and when darkness comes and hides their dark
deeds they destroy all our plants. If you do not
plant vegetables the cut-worm is not at all loath to
eat your pinks and asters.

They are thick, dark-colored, disagreeable looking caterpillars of a dark lead color (Figs. 139 and 141). Their chrysalides are of a bright mahogany color and the moths come out between the middle of July and the middle of August. There are a number of species of these moths, but the principal difference among the caterpillars themselves is the difference in the degrees of their meanness.

Seriously, however, boys, if you will collect a number of these cut-worms from the soil of your flower garden, your kitchen garden, your potato or corn patch, along in June or July, and put them in the boxes of earth, they will probably, everyone of them, immediately conceal themselves in the dirt and soon change into pupa or chrysalis form and when the moths break out of their mummy cases you will find you have many different kinds, although, to the careless observer, the " worms " looked all alike.

CLOTHES MOTHS

This name is used for several different moths, the larvæ or young of which eat woollen clothes, furs and feathers, and like the basket caterpillar on our trees, and the caddice worms in our streams, they use the material upon which they feed to build themselves houses carried, after the manner of snail shells, on their backs. Fig. 143 is a very much enlarged view of the young clothes moth which you will find eating the woollen clothes packed away in dark closets—not the moth itself (Fig. 145) you must remember, but it is the larva that does the damage. Fig. 144 shows the chrysalis state of this pest.

The moths or butterflies are very small and of a light buff color, with a shiny silk lustre; the

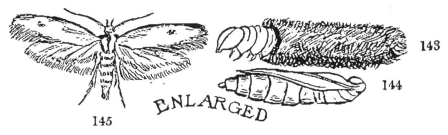

body part or abdomen is paler than the front wings and the hind-wings are also lighter in color. It has a luxuriant head of blonde hair and its wings are long and fringed most beautifully with blonde silk. When we say long wings, we do not mean that they are great in dimension, but they are long in proportion to their width.

You will find the moth flying about the house in the latter part of April and the first part of May, and when you seek to destroy it you will discover that you have something to learn in the art of hunting moths. These little things apparently have learned from experience exactly how to dodge and evade the human hands. When one thinks one has certainly got the little butterfly between the palms of one's hands and brings the latter together with a resounding slap, it is only to find that, like a tumbler pigeon, the moth has dropped beneath the danger zone. One is lucky if one catches a fleeting glimpse of the pursued as it zigzags away, using the tactics of an Indian dodging rifle fire, and disappears in a dark corner where search for it will be in vain. That is, it will be in vain until one has learned the tricks of this little enemy, then one will look for a crevice or crack at the spot where it disappeared, and probably with a knife blade inserted may bring the criminal to light and well-deserved execution.

CONCLUSION OF THE MOTH TALKS

If the reader desires to make a scientific study of moths, he should make his collection of caterpillars, of chrysalides and cocoons as well as the moths themselves. By collecting a number of the caterpillars and preserving some in spirits, being careful to number the vials containing them and

allowing the others of the same kind to go into
cocoon state and preserving some of the chrysalides
and cocoons in spirits, numbering them the same as
the caterpillar, also allowing some of the caterpillars to hatch out as moths and preserving a specimen of both male and female millers and numbering them the same as the caterpillar, the reader
will have the data necessary to completely identify
his specimens, and if he adds the eggs of the moth
on the leaf, stick or bark upon which they are laid
and preserves them in spirit, he will have the whole
life history of his specimens.

When he makes a collection of this kind and
goes into it scientifically he should secure Packard's
Introduction to the Study of Insects, and read that.
Of course he must expect to have many bruises
from knocking against the hard names he finds in
this book, but after a while his mind will become
toughened and contact with the hard names will
cease to pain him. He should also hunt up a copy
of Harris's Insects Injurious to Vegetation and
read works by such men as Leland O. Howard
and J. H. Comstock, and such women as Ida M.
Eliot and Caroline Gray Soule.

Scientific books on moths and butterflies are to

be found in all our big libraries, and where libraries do not exist the reader may send duplicates of his specimens to the scientific men of the State, City or National Museum. They are all good fellows and will gladly identify the insects for him and spell out the long names so that he may label them properly.

These names are purposely omitted here because this is a book for boys, a book the purpose of which is to interest the reader in the study of nature and not to frighten him out of the glorious fields and away from the enjoyment of the sunshine and blue sky by building up barbed-wire entanglements of long Latin and Greek names.

The reader may even become an authority on the life and habits of bugs, butterflies and beetles without even knowing one scientific name; afterwards, when he is older, or whenever he feels like it, he may gradually acquire such a knowledge of the scientific names as will make the other boys speak of him with bated breath and look upon him with awe!

CHAPTER TEN

BUTTERFLIES

You remember, back in the part of this book where we were talking about the moths, mention was made of the beauty spot, or nature's conspicuous decoration on a lot of the millers. Some of the butterflies also have this trade-mark (Fig. 146) and among the swallow-tails you will find it on the inner edge of their lower wings, just at the top edge of the border band. It would be very interesting to know why nature is so very fond of this beauty mark, but it would have been a knotty problem for Huxley, hard for Darwin and difficult for Mr. Wallace, or any of the rest of the evolutionists, to give us a satisfactory explanation of the reason that this same decoration appears upon birds, fish, insects and mammals.

For those people who do not believe in evolution and who think things just happened, without any long series of education, preparation and gradual growth—for such people, with such beliefs, it might be easier to account for the beauty spot, because such persons do not have to explain, they can simply say, " It is there because it is there." But persons who are gifted with a healthy

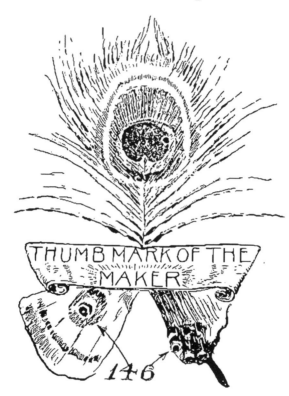

imagination, such as Indians, artists, story writers, poets and boys, need not prove their assertions, they can simply make the claim that these round eye-marks which appear on so many creatures are the thumb-marks of the Great Creator, and in one sense they are, no matter how they may be otherwise explained by Darwin. At any rate we can all agree that the Creator could hardly have chosen a more beautiful and artistic object upon which to register His thumb-marks than a swallow-tailed butterfly or a Luna moth.

It is one thing to have a thought, and it is an entirely different thing to so express that thought that another person can understand it. Now listen to me a moment, boys, and see if you can understand what I am going to tell you; see if you can catch the thought which underlies the whole of this book, by which I mean the main idea that governs me in writing this book. A number of times I have joked about scientists and, while I greatly respect them, for my own purposes here I am going to joke about them again. The particular scientists I mean are the postage-stamp naturalists, those devoted students who spend all their time classifying and sorting out dried bugs, butterflies and beetles, indexing them and preserving them in order, as one does a collection of postage-stamps in an album.

These men do not and cannot tell you what a butterfly really is, they cannot tell you why it exists, why it lives at all! They cannot tell you what impulse, thought, instinct or motive governs its actions; in fact, they can do no more than guess at these things, as we do; they cannot answer one real live question which any bright boy would ask regarding the life of a butterfly.

Of course they can give you a name from a language as dead as the specimens in your collection; they can also tell you that the specimen was once a caterpillar, and sometimes, not always, can tell you what kind of a caterpillar. Frequently they can tell you what sort of eggs the butterfly lays, where it lays them, how long it takes them to hatch, etc. They will also tell you that the caterpillar changes its skin about four times in its lifetime; that after a while it stops eating and changes into the form of a chrysalis and then becomes a butterfly. But if they are real postage-stamp scientists, you will not know what they are talking about because the terms they use are seldom heard in conversation and do not appear in print except in dictionaries and scientific reports.

But even if you understand what they say, you are still as ignorant as they are of the meaning of a butterfly! The specimen in your collection is of the same value as a postage-stamp in an album.

Your specimen is not a real butterfly and it bears no more relationship to the live insect than does an Egyptian mummy to a life-loving, rollicking wild cowboy.

A butterfly is something more than an insect, it is an idea,—it is everything that you want it to be, and it is beautiful in proportion to your ability to appreciate and understand beauty.

What is the object of its life? What is it for? These are questions which come to the mind of any healthy boy; stupid men seldom think of them, even when they see one of these exquisite insects floating in the air apparently as aimlessly as a piece of tissue paper wafted on the summer breeze. Why, boys, if we knew all the hidden secrets of the life of one single butterfly we would know more than any man who ever lived. But there are lots of secrets we *can* discover and that is what gives charm to our collecting hikes.

This book is written as an appeal to the sensitive nature of you boys—boys whose souls are nearer to nature and whose spiritual ideality, if you can understand what that means, is greater than that of men, at least greater than that of most men. There was once a man by the name of

Thoreau who possessed ideality, which every real boy has. Thoreau had a warm fellow-feeling and real sympathy for everything that lived, and the joyous enthusiasm of a boy because he had the clean soul of a boy.

You, my readers, are all Thoreaus because you are BOYS. And it is because you are boys I write as I feel and not as some men would have me write of the butterflies we see glinting in the sunlight, flitting from flower to flower, idly loafing on a milkweed blossom, opening and closing their wings in their dainty, languid fashion, or collecting in crowds and making blotches of moving color around the damp places in the roads and barnyards.

Yes, butterflies are beautiful, they are artistic, but there is another side to the story: they are the good Dr. Jekylls of that famous novel and the caterpillars are the wicked Mr. Hydes.

Everyone who is interested in forest shade trees, in farms, in flowers, in gardens, and everyone who thinks he is not interested in these things, but uses wooden furniture made from forest trees, eats vegetables, fruit and grain grown on the farms, wears a flower in his buttonhole, uses paper on which to write, is depending for all these things upon what the caterpillar spares only through lack of numbers to consume.

We eat at the second table after the caterpillars, we use what is left after his majesty the caterpillar has had enough. Consequently, everybody in this world, whether he knows it or not, is dependent upon the birds, principally; also, upon some of the bats, toads, small snakes and some small mammals which eat insects, for the privilege of living here.

Although the caterpillars are baby butterflies, that fact need not disturb you when you are waging war against them. Every time you catch a female butterfly and put her in your cabinet, you have cut off just so many hundred eggs, which means so many caterpillars from the general supply of marauders.

Nevertheless the caterpillar has its use and place in the world and should not be exterminated. However, this need not bother you because *you cannot exterminate them,* but unless you keep them in subjection they will exterminate you! I am telling you this, not to encourage you to kill, but so that you may collect specimens with no scruples of conscience; although the real nature lover dislikes to kill, even pests.

CATERPILLARS

It is the common belief that caterpillars are always hairy, but we have seen that among the caterpillars of the moths many of them are naked, and such is also the case with the butterflies. Whenever we use the word butterfly now, we mean the

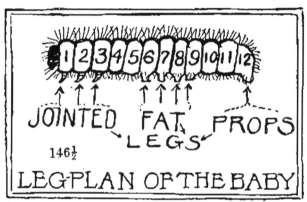

ones which fly by day, and not the millers. The caterpillars which turn into butterflies always have sixteen legs. They have a pair of scaly-jointed legs attached to each of the first three divisions of the body and they have four pairs of fat fleshy legs attached to the divisions, 6, 7, 8 and 9 of the body (Fig. 146½). These fat legs have no joints at all to them; they are shaped something like piano stools. Besides the fat ones they have a pair of prop legs on the last section of the body. All these fleshy legs are soft and can shape themselves to fit the branch upon which they rest. The feet to them, if we may be allowed to use such a word for the bottoms of these fat legs, are nothing more than cushions. That they are not real legs, but only artificial limbs to help the baby butterfly creep, is shown by the fact that when the insect comes out of the chrysalis in perfect form, there are no props on the tail and no cushioned piano stools along the belly.

The caterpillars to the butterflies have a habit of hanging themselves by their tail before changing to the chrysalis form, or putting a belt strap on around the body, so that after they have shed the caterpillar skin and becomes a helpless mummy the band of silk holds them to the limb of the tree or the paling of the fence until the butterfly emerges.

The butterfly is much shorter, as a rule, than the miller and it also is more slender and graceful in its body and of a livelier disposition than its night-flying relative. It disdains a cocoon and delights in artistic and decorative mummy cases in which secretly to change its clothes. It, too, has a

spiral tongue (wound like a watch spring). When
at rest, the butterfly holds its wings upright, erect,
never folds them as do a great many moths. No
bristle and socket can be found on the butterflies'
wings to hook them together during flight, like a
woman's dress, as the wings of some moths are held
together. Most of the smellers, feelers or antennæ
of the butterflies are knobbed on the end, although
some approach very closely to the thread-like form
which naturalists call " filiform."

THE SWALLOW-TAILED BUTTERFLIES

The most attractive butterflies are the swallow-
tails; the so-called swallow-tails to the wings of the
butterfly are marks of distinction, as they are on the
wings of the Luna moth. There are over three hundred kinds of swallow-tail butterflies known; the
three hundred does not refer to the number of butterflies, because one may see that many in a day.
Butterflies sometimes migrate in great flocks and I
have seen them in clouds, floating over the house-
tops of New York City. On such occasions one is
liable to see many, many times three hundred in
one day.

Among the butterflies, so far as I know, there

are no wingless ones, but among the moths there
are some of the females which never have wings;
they are the ones to which the following misquota-
tion may apply:

I do not want to fly, she said,
I only want to squirm,
I hate to be a butterfly,
I want to be a worm.

Such moths lack ambition, but unlike many
females they cannot be accused of vanity, they do
not care for powder or paint or perfume, and that's
where they differ from the black swallow-tail butterfly or, as some call it, the swallow-tail papilio,
because this butterfly is powdered and painted and
uses perfume. The perfume is used only by the
caterpillar. The baby swallow-tail has " eyes and
sees not," six of them on each side of the head at
that!

It has strong jaws which open and shut side-
wise and in the lower lip there is a little tube from
which the silk or web threads are drawn or, as they
usually say, spun. This silk when it comes out,
you understand, is not in the form of thread, but a
sticky sort of juice and it drools out of the lower

lip in a small thin jet which the air hardens into the substance we call silk.

The caterpillar of the black swallow-tail has peculiarities all its own. You possibly know it as the parsley " worm "—I do not know what we are going to do about that word *worm* which comes up on all occasions; I suppose we must use it as other people use it, but if while doing so we understand that it is incorrect, that in reality it is a slang word for caterpillar, it will probably do us no harm— the caterpillar may be found in June, eating the leaves of the carrot and parsley. It is a naked larva of yellow or green color, striped and spotted with black markings.

If you touch the parsley " worm " it will defend itself by protruding, from a slit in the first division of the body, a delicate pair of soft orange-colored horns which are joined together at the bottom, making the letter V. The caterpillar will not gore you with these horns, you can touch them with your finger without injury to yourself; the truth is, they are not real horns and they are only called horns because of their position and appearance.

The V-shaped thing over the caterpillar's head is really its vinaigrette, its perfume bottle. This is another case where the insect's idea of a sweet odor does not agree with ours; but maybe it is, as we first hinted, used like a skunk's odor, as a means of defense.

ICHNEUMONS

There are a number of flies, and what are called Ichneumons, which have a very annoying and mean way of depositing their eggs upon the surface or under the skin of caterpillars, where the eggs hatch out and feed upon the flesh of their living host. Possibly this vinaigrette carried by the striped caterpillar is used to drive away all such insects as wish to pasture their young upon the body of the live caterpillar, or the smell may even be so disagreeable to the toads and the birds as to cause them to refuse to eat the caterpillars. We know the scent is there for some purpose and we know we would not eat one of these caterpillars even if it had no vinaigrette bottles stowed away in a pocket in the nape of its neck, and we also know that when we see a woman bring out a vinaigrette bottle we must not mention parsley " worms," for that would be ungallant.

These caterpillars are full grown in the fore-part of July and will then measure about an inch and one-half in length. This is the time they hunt a sheltered spot on a tree trunk, shed or fence, there to get ready to effect their wonderful transformation. They are dainty, fastidious creatures and they want a footstool for their feet, so they make themselves one of silk, but instead of standing upon this silken stool, they hook their hind feet into it, fasten them to the silk so that they can, if they wish, hang head downward with no danger of falling. But they evidently do not like to hang head downward, and in order to avoid that undignified position they spin a waistband or lifebelt, which keeps them upright and prevents the blood from running to their head as you will notice in the picture of the black swallow-tail.

These butterflies not having a cocoon, like the moth, to conceal their chrysalides, take some pride in their mummy cases and make them of decorative and artistic form to please the eyes of the boys or for some purpose of their own.

The butterfly is black and is common; every boy knows it, or if he does not, every boy has seen it. It is graceful in form and beautiful in color. The wings have two rows of yellow dots and a lot of yellow half-moons along the border of the wings,

the half-moons being known as lunules. The male
butterfly is more distinctly marked than the female,
and also smaller.

You can put it down as a rule among all insects
that the male is smaller than the female; the exceptions, if any, which may occur to this rule are not

BLACK SWALLOW-
TAIL

many or important enough to affect the general
truth of the statement.

THE GREEN-CLOUDED SWALLOW-TAIL

It is also a black butterfly with yellow markings,
but if you will compare it with the black swallow-
tail you will see the difference in its markings,

especially in the fore-wings. If the drawing was in color you would see that the bands on the hind wings are not yellow but, as the name indicates, clouded with green.

151

THE TIGER SWALLOW-TAIL

This is the butterfly we formerly called the turner, a corruption of *turnus,* but since it is yellow and striped with black, the name tiger is more appropriate. This is a big, handsome, conspicuous butterfly, expanding sometimes as much as five inches across the wings.

The caterpillar you will find feeding on the

leaves of the wild cherry and apple trees. The full grown caterpillar is sometimes two and a half inches long. It has yellow eyespots with black centres on each side of the third ring of the body. The upper part is of green color with rows of little blue dots and there is a yellow and black band

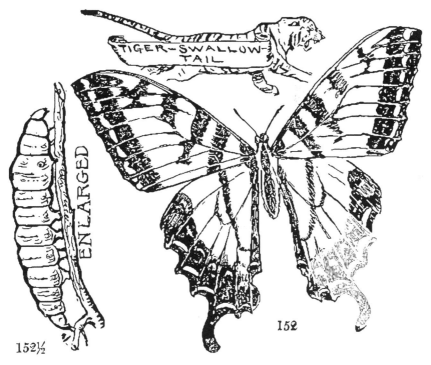

across the fourth division of the body; it wears fashionable pink stockings.

You will find the chrysalides about the first of August, but you will have to keep them until the following June before they hatch out butterflies. The larvæ of all these butterflies carry vinaigrettes with ill-smelling perfume and probably the larva of the zebra butterfly does the same, but I am not familiar with this caterpillar, although the butterfly itself is an old friend.

THE ZEBRA SWALLOW-TAIL

We used to find this in the edge of the woods

and among the boys it was known as the wood butterfly. There are different forms to this insect and when referring to your notes you may notice a difference in the time of the butterflies' appearance. Their wings have stripes of greenish white, which gives them the name of the zebra swallow-tail.

ROUND-WINGS

The next butterflies are the round-wings. They have short thick antennæ with a rounded club at the end and the point of the fore-wings is rounded off. They are mountaineers, and after these mountaineers come the inhabitants of the valleys, some familiar inhabitants we all know.

CHAPTER ELEVEN

WHITE CABBAGE BUTTERFLY, YELLOW BUTTERFLY, THE GOSSAMERS, COPPER AND BLUE GOSSAMERS, THE MONARCH BUTTERFLY, THE VICEROY BUTTERFLY, THE APHRODITE AND MYRINA BUTTERFLIES. THE PHAETON BUTTERFLY, ANGEL-WING BUTTERFLIES, THE L BUTTERFLY, THE AN-TI'O-PA BUTTERFLY, THE RED ADMIRAL, THE BROWNIES AND THE SKIPPER BUTTERFLIES.

BUTTERFLIES

EVERY lad who has hunted butterflies is familiar with the white cabbage butterfly, which may often be seen in great numbers in the cabbage and turnip patches; some of them have dusky tips to their wings with a few dusky spots upon them, while others are white with dusky color near the body. The accompanying illustration shows the white cabbage butterfly, the green larva and the chrysalis.

The caterpillar is covered with dense hair and is of a dull-green color. Some of these caterpillars which I kept in captivity were devoured by a lawless Isabella caterpillar confined in the same box. By turning back to Fig. 123 you will see a sketch of this cannibal.

THE YELLOWS

After the whites come the yellows, which are almost, if not quite, as common as the white butterfly. These, you will notice, like the whites, have no tails to their wings and both the fore and hind wings are more rounded and have smoother edges than those of the swallow-tail butterfly.

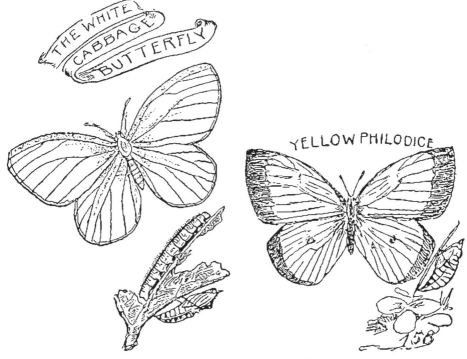

There are two broods of yellows every year, the first coming in April or May, the last in July. The female butterfly, in the latter part of this month, deposits its eggs which hatch out about the first of August. The minute hairs on the body of the caterpillar give it a downy appearance, and it has yellowish white stripes on each side of its body. They feed on clover and green-pea vines. The chrysalis has a belt of silk, like those already described, and the head of the chrysalis is pointed. These yellow butterflies are common as far north as bleak Labrador and our own country roads in summer time would not look natural without them.

There are also, by the roadsides, in the fields,

147
BRONZE

some very small butterflies which will attract attention on account of their dainty appearance, known as gossamers. They include the coppers and blues.

THE AMERICAN COPPER BUTTERFLY

The American copper butterfly (Fig. 147) is easily recognized by the red copperish sheen on its fore-wings and the eight, more or less, small square black spots. The hind-wings have a broad dusky brown border and a wide copper-red band on the back margin. The butterfly spreads a trifle over an inch. You will find him among the clover and pasture plants. The larva is a greenish-colored caterpillar and the chrysalis (Fig. 147) is short and dumpy in appearance, yellowish-brown in color and peppered with small black spots.

The blue butterfly (Fig. 148) is a most attractive little fellow and very beautiful. It will spread its wings about the same distance as does the c o p p e r butterfly. The wings have a satiny lustre and in the male butterfly are an azure blue color; the female has fore-wings with w i d e, dusky outer margins and she has a row of black spots on her hind-wings. The under sides of her wings are pearl gray and the fringes are white.

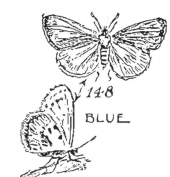

148
BLUE

If you hunt for them you will find other coppers and other blues—the Blue Lucia, for instance.

THE MONARCH

But we must skip a number of these dainty little fairies and take our butterfly net out along the fences and roads in search of royal game. There

we will undoubtedly find the Monarch or milkweed butterfly. This butterfly has a pretty chrysalis of bright-green color ornamented with golden
beads; you will find it sheltered under the project-
ing top of the old white board garden fence.

The caterpillar feeds on the different kinds of

THE MONARCH

milkweed. It is yellow in color and has broad
bands of black. There is also a pair of thread-like
appendages growing on the second division of its
body and another pair on the eleventh division.
The butterflies are very common and, so far as I
know, do no injury to any of the garden plants or
vegetables and not any serious damage to the milkweed upon which they feed.

THE VICEROY

There is another butterfly very much like the Monarch, known as the Viceroy (Fig. 150). It is the same color as the Monarch, but is smaller and differently marked, the principal difference in marking being the band on the hind wings; but although these two butterflies look so much alike, their resemblance is not due to close relationship, for the scientists have declared that they belong to different sub-families— that is, they are about fourth cousins to each other. The markings and color, however, are very much alike. The Viceroy, like the Monarch, is a tawny yellow above

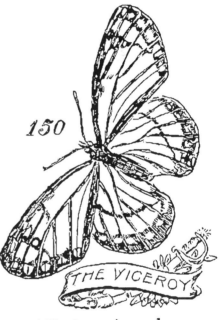

and a paler yellow beneath. All the wings have a wide black border relieved by a white spot, the veins of the wings are black and there are triangular-shaped spaces with white spots near the tip of the front wings. This butterfly can spread about three and one-half inches. The light-brown caterpillar feeds upon the willow and poplar leaves; the young caterpillars have an ingenious way of making themselves sleeping-bags by neatly joining the opposite margins of a willow leaf, lining the bag with silk and sleeping in it all winter.

THE APHRODITE

The Aphrodite (Fig. 151) is a double-brooding butterfly, the first specimens of which you will

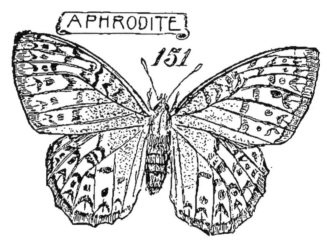

discover about the middle of June, and new Aphrodites fresh from their chrysalides may be found in the latter part of August. It has tawny yellow wings—that is, the males have—while the females have what might be called ochre-yellow, and both gentleman and lady are of a brownish color next to the body and near the hind edges they have a black line. A row of black new moons and black full moons on the other part of the wings are ornamented with irregular black spots. The Aphrodite is not in favor of a gold standard, but on the contrary is a free-silver butterfly and beneath the tips of the front wings it carries seven or eight silver marks, while concealed on the under side of the hind-wings are twenty-odd great, silvery-white spots. You must look for this butterfly among the blossoms in the lowlands—it is not a highlander.

THE MYRINA BUTTERFLY

The eggs of the Myrina butterfly are about the shape of an acorn and pale green in color. The young are hatched in about a week's time and are full-grown at the end of the first week in August.

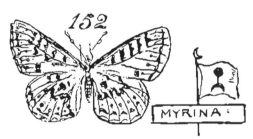

The head is black and shiny and coated with fine, short black hairs; the sort of grayish-brown body is ornamented with spots and dots of black velvet. The second segment or division (Fig. 146½) is ornamented with two fleshy horns; the third and fourth divisions each have dull white colored spines with black tips; all the other divisions, except the last one, have six dull white spines; there are four of them on the last or twelfth division. It wears black patent-leather shoes on its front feet and tan-colored ones on its fat legs. Fig. 146½ shows the location of the fat legs.

The butterfly has tawny wings with black border and a row of black new moons next to the border. It spreads less than two inches across the wings. Mr. Harris gives the figures as from one and three-quarters to one and eight-tenths inches. It, too, belongs to the free-silver party and is ornamented with silvery spots as well as black dots (Fig. 152).

THE PHAETON BUTTERFLY

The Phaeton butterfly (Fig. 153) you must hunt for in the swales and over damp soggy ground. You can also look for the caterpillars in the spring, quite early, and maybe under the leaves you will find them hiding. The full-grown caterpillar (Fig. 154) is armed with nine rows of black spines surrounded at the tips with thick-set long spinules. The caterpillar is ready for a minstrel show, for it has a black face, the front part of its body is also black, but the rest of the body is clothed with an orange-colored garment. Along about the first of June you may find the chrysalis (Fig. 155).

THE ANGEL-WING BUTTERFLIES

This is a pretty name which I quote from Mr. J. H. Comstock. It is a pity that more of our butterflies are not named in this style, but at the same time, according to the best of our information, it is not the angels but the fairies who s p o r t butterfly wings. We may be wrong in this

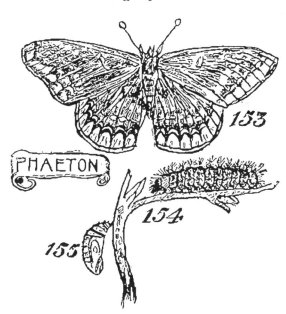

because, to be honest, we have never, to our knowledge, seen either of them.

But as an artist the author has many times drawn pictures of angels and, taking his authority from other artists, he has always hitched birds' wings under the angel's shoulder-blades, not because he thought angels needed wings but because the wings are decorative and symbols of the angel's ability to move through space. But this point we will not discuss because the writer of Bugs, Butterflies and Beetles is more familiar with boys than he is with angels, and, fortunately for school-teachers, policemen and parents, boys do not have wings.

The peculiarity of the angel butterflies seems to be that when Mother Nature was using her shears to cut out their wings, she made many experiments and gave these butterflies all sorts of fancy notches, scallops and curves on the edges of their wings. The scientists would say that Mother Nature gave them " deeply incised wings."

The angel-wings are also painted with rich reds and browns and usually they have the under side of their hind-wings decorated with silver and gold spangles. It may be, in order to help you boys remember how to indicate the stops and pauses when writing your notes, that these butterflies often have their wings ornamented with punctuation marks. One of them has a golden semi-colon and one angel-wing is called the question-mark butterfly or, to state it more accurately, the interrogation butterfly. It is a rich, reddish-brown color, with fancy notched and tailed edges to its wings,

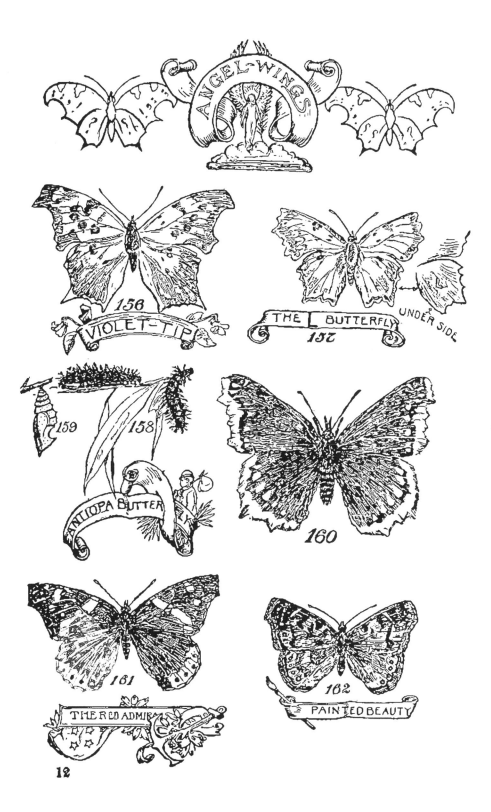

ANGEL-WINGS

156 VIOLET-TIP

THE BUTTERFLY 157 UNDER SIDE

159 158

ANTIOPA BUTTER

160

161 THE RED ADMIRAL

162 PAINTED BEAUTY

12

which are tipped with violet and marked with dark spots. The caterpillar may be found on the elm trees, the hop vine and the nettles (Fig. 156).

THE L BUTTERFLY

The L butterfly is so-called because branded like a Western broncho—that is, it carries a silver L in the middle of the under side of its hind-wings. The caterpillars of the L thrive on the leaves of the hop and the elm trees. The L butterfly (Fig. 157), is a northern variety. Further south we have a comma butterfly, which is branded with a silver comma in the centre of the hind-wings; the caterpillar of the comma also feeds upon the hop and elm trees and the nettle.

THE ANTIOPA

The Anti'o-pa is a hyphenated American and not a native-born citizen of our republic, but like all the rest of the immigrants, including our own far-distant ancestors, the Antiopa came over here to better its condition and found here fewer enemies and plenty of food and so it has thrived like the rest of the immigrants, and become one of the citizens of our butterfly community. The caterpillars play hob with the willow trees. Some weep-ing willows, which with great trouble and some little expense I procured and set out along the edge of Big Tink Pond, Pike County, Pennsylvania, were completely stripped of their leaves by the larvæ of this imported butterfly.

The greedy babies are black, lively caterpillars and they live together in numerous communities, the first brood coming early in June—that's the time they began on my willow trees, and two seasons' diligent work by these caterpillars killed every tree I had. They have black heads, with spines sticking up from them. They have six or seven of these jagged spines on each division of the body. When full-grown they are an inch and three-quarters long, and they do not look at all pretty; in fact in olden times they were supposed to be very poisonous and able to give you dangerous wounds and they certainly look like villains. At one time, people cut down all the poplar trees around their dwellings because they were afraid of the Antiopa caterpillars, which feed upon the poplar as well as the willow. Fig. 158 shows the caterpillar, Fig. 159 shows the chrysalis and Fig. 160 the butterfly.

I have found the butterfly in mid-winter under-neath stones which were half buried in the frozen ground. The butterflies in the fall creep edge-wise in the crevices leading underneath these rocks, and sleep there all winter so that they are usually the first butterfly one sees in the spring. A real warm spell in winter time will sometimes induce them to come forth and flit around in the sunshine, under the belief that winter is over. The butter-fly wings are dark purplish-brown above, with the band along the scalloped margin of buff color. Ad-joining, or rather just beyond, the buff edge is a row of bluish spots. The butterfly spreads about three and one-half inches at most.

THE RED ADMIRAL.

This is another angel-wing, the caterpillars of which feed on the nettles and hops (Fig. 161). After the Red Admiral comes the cosmopolitan Painted Beauty. This butterfly is right up to date so far as paint and powder are concerned, but if she does the turkey trot or the tango, she does them while flitting through the air and without a partner. The Painted Beauty (Fig. 162) in color is very much like the Red Admiral, although the markings are different, which you may see by comparing the two sketches. Like the donkeys, the caterpillars of these butterflies all seem to love nettles; this is true also of the American Tortoise Shell. There are two Tortoise Shell butterflies, the Compton Tortoise and the American Tortoise.

THE BROWNIES

These butterflies have the eye-spots or the "thumb mark of the Maker" as their favorite decoration; they are sometimes called the meadow-browns, because they frequent the meadows.

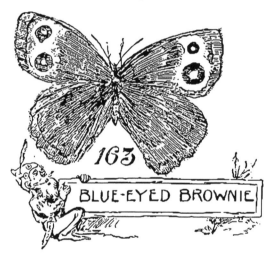

163

BLUE-EYED BROWNIE

The Blue-eyed Brownie (Fig. 163) may be found about the first of July to the middle of September in the orchards and woods. The dark-green striped and pale-green body of the caterpillar changes to the chrysalis form with a notched head; the front wings of the butterfly have a wide yellow band near the outside edge and extending to the middle of the wing or further. At the top and bottom of this yellow blotch of color are two eye-spots with blue centres. The hind-wing is more or less scalloped and the under side of the wings is of light-brown color, streaked with dark brown and ornamented with eye-spots or nature's beauty spots on the females, but not always on the males. Up North these butterflies will measure two and one-half inches across the wings and a half an inch more for the South.

BOISDUVAL'S BROWNIES

These butterflies (Fig. 164) are a pale yellowish brown. Both sides of the front wings have a row composed of four eye-marks. The eye-spots are black with w h i t e centres. On the back or hind wings there are six eye-spots, one of them on the upper edge of each wing and five of them close together on the

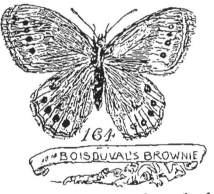

164

BOISDUVAL'S BROWNIE

lower edge of each wing. It is not unusual to find some of these butterflies with blind eye-spots upon the upper side of the wings—that is, eye-spots lacking the white centres.

The butterfly spreads two inches and sometimes more. It may be found in July among the mountain meadows and on the hillsides in New England. It was named by Dr. Harris after Dr. Boisduval, the entomologist.

There are other Brownie butterflies, which you will know and the tribe they belong to because of their family likeness. There is, for instance, the Eurytris Brownie and the Nephele Brownie. But we have given enough space to the Brownies and to tell the truth, they do not look as much like the little gnomes for which they are named as do the Skippers. The Brownies are called Brownies because of their color and not because of their habits or form.

SKIPPERS

The Skippers, however, have all the characteristics of little dwarfs—big heads, bulging eyes, and short heavy-set bodies (Fig. 165). Even the baby Skippers, the caterpillars (Fig. 166) have big heads. These caterpillars are leaf rollers. While making this illustration, I was unable to find on the locust trees the larva of the Tityrus Skipper (Fig. 165), but I found a leaf roller on a silver poplar tree (Fig. 167), which will serve as an illustration of the ingenious manner in which leaf-rolling caterpillars roll up the leaves.

At *A*, *B*, and *C* you will note (Fig. 167) that
the roll is fastened by stitches, if I may use that
term, of silk. These stitches continue at intervals
inside the leaf as it is rolled, thus holding it together in the form of a tube. Inside the tube the
caterpillar leads a hermit life, concealed from its
enemies by its food supply. This particular caterpillar feeds upon the edge of the silver poplar leaf

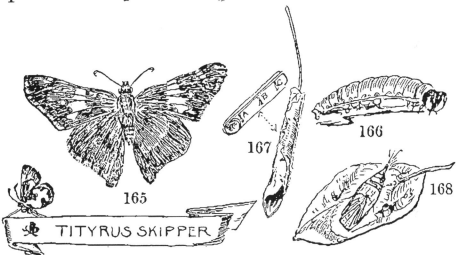

166
167
165
168
TITYRUS SKIPPER

inside the roll. But when one unrolls a leaf one
finds the caterpillar to be an unsanitary housekeeper. The larvæ of the Tityrus Skipper, however, is the reverse of slovenly. The Tityrus keeps
one end of the leaf roll open as a doorway (Fig.
167), from which it is said to come out at night,
feed and return to its hiding place when the sun
rises and exposes it to the view of its enemies, the

birds. The Tityrus Skippers are good housekeepers; they have no dirt in their bedrooms and have a way of throwing it out, by jerking their body and casting the refuse quite a distance. The caterpillar of the Tityrus feeds upon the locust trees and sometimes makes its cocoon inside the leaf which it inhabits (Fig. 168). But usually it seeks some safer place and makes its cocoon of any old loose stuff it can find, lines it with a web of silk and sleeps there until the following summer.

The butterflies, the real true butterflies, when at rest, bring their wings together like the leaves of a book, holding them stiffly upright in this position. But some of the Skippers bring their fore-wings together upright like a butterfly, while holding their hind wings partially open like a moth or miller when at rest. Other Skippers make no pretense to holding their wings upright, but spread them open like the moth when at rest.

They also have a tendency to make cocoons like a moth's instead of suspending themselves in jewel chrysalides, like the real butterflies.

The Skippers' bodies are thick and suggest the bodies of the moths more than they do those of the butterflies. Then you will note that their antennæ are very much like the antennæ of the Sphinxes or hawk moths, and for lack of any rule to the contrary we will consider them the "missing link" between the true butterfly and the miller. If they are not, then the link is really missing.

The only serious objection to butterflies, as objects of study, is the difficulty in keeping them alive. When one confines them in the house, they have a foolish way of fluttering on the window pane or beating their beautiful wings to rags on the window screens. They make splendid objects to preserve and are beautiful in form, texture and color; they add to the beauty of a collection and when alive add to the sentiment and beauty of the pasture, the meadow and the garden as they flutter in the air, but their children's energies are all expended in an effort to destroy the beauties of nature. As caterpillars, nothing has a value to them but food for themselves.

We cannot keep butterflies in the greenhouse unless we are careful to secure only males, otherwise the insects will deposit eggs upon our plants and transform the greenhouse into a caterpillar farm.

But when we come to the next sub-division of

our book, the beetles, ah! that is a different proposition because one may keep these alive for an indefinite length of time in boxes and cages made for them.

But before closing the chapter on butterflies, let me tell of the butterfly, a storm-beaten individual, I found hapless and helpless where the rain had beaten it down by the roadside in Connecticut last season. It could not fly because its wings were uneven, I felt sorry for it in its pitiable condition and I placed it upon the railing of a fence, and taking a sharp blade of my pocket knife, trimmed both wings off nicely and evenly, making them each exactly the same size, although much smaller than they were originally. Then I released the butterfly and was pleased to see it fly away as easily and apparently as care-free and happy as if nothing had happened.

From a sentimental point of view this was a very pretty incident, and the novel first-aid work rendered to an injured butterfly will appeal to the sympathies of all tender-hearted people. But the practical results of setting that butterfly free might be the establishment of a colony of voracious caterpillars. The experiment, however, was interesting, and I trust the results did no harm to the farmers.

I have kept grasshoppers, katydids and other interesting specimens alive in the house until after the winter holidays. The katydid was fed on lettuce and was a most comical and amusing pet. It met its death by creeping into the ashes of the open fireplace and not getting out of the way when the maid built the fire New Year's morning. That's what Katy did!

CHAPTER TWELVE

COLEOPTERA. NAMES OF PARTS OF A BEETLE. GRUB-WORMS AND WHERE AND HOW TO COLLECT BEETLES. LIVING SUBMARINES AND HYDROPLANES. A DOODLE TRAP. PET BEETLES. WHIRLIGIGS. LIONS AND TIGERS OF THE PONDS. HOW DIVERS CARRY AIR UNDER WATER.

WE now come to that numerous tribe called the "beetles." To the writer's mind they are more interesting to the boys than any other race of insects. They possess certain characteristics which appeal to boys, by which I mean they have certain things about them which make them good playmates for boys. In the first place, none of them are poisonous; in the next place, none of them will hurt a boy who knows how to handle them, and in the third place they are as a rule so stoutly built and so thoroughly armored that, with ordinary care in handling, there is little danger of injuring the insect itself by playing with it. Added to this they are often very comical.

Bugs are unpleasant to handle; wasps, bees and hornets are, to say the least, very inconvenient things to handle. They are hot-tempered and have a hot needle with which they puncture the skins of

190

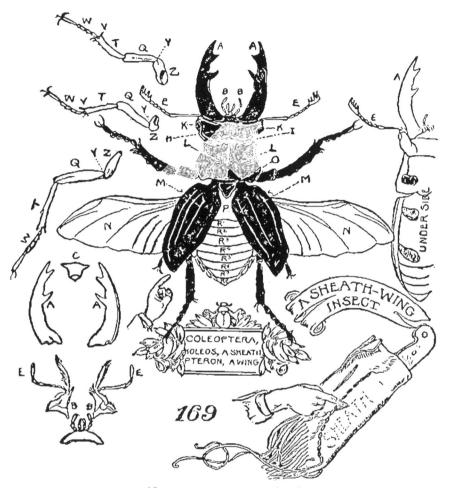

Names of the parts of a beetle.

A, jaw bones, pinchers (mandibles); B, one of the small feelers (palpus); C, lip (labrum); E, the big feelers (antennæ); H, back of the head (occiput); I, neck; K, eye; L, chest (prothorax); M, wing cover (elytron); N, hind wings or back wings (front wings are hardened into wing covers); O, shield (scutellum); P, outside of the back of the last part of the thorax, metathorax (metanotum); Q, the thigh or upper part of the leg; R, R, R, rings of the belly or abdomen (tergites); T, shin-bone (tibia); V, spurs; W, feet (tarsi); Y, hip-joint (trochanter); Z, hip bone (coxa).

their captors. Butterflies and millers are far too
delicate to handle, but beetles, with the possible
exception of the carrion beetles and the soft-bodied
oil beetles, possess none of these disadvantages.

Beetles are six-legged insects, and, with few
exceptions, they have a pair of thick, horny front
wings which are of no use while flying, but when at
rest act as covers for the hind-wings, fitting together like the shell of a turtle. Beetles also have
mouth parts for biting and chewing.

Beetles, like butterflies, start as a worm-like
creature, then go into the mummy state, from which
they emerge as beetles. Fig. 169 shows a beetle
as an insect with six legs; it also shows the wings
extended and the fore-wings, which form the
sheath and give the name to the family, are spread
apart. To the right or east of the beetle is a sketch
showing the under side of one from which the legs
have been removed. To the west or the left are
the legs, in the southwest corner of the drawing
are the mouth parts, in the southeast corner is a
sketch of my hunting knife in its sheath; this is to
show what a sheath is. The knife I thrust in the
sheath from the top down, the beetle folds its wings
over its body then shuts its sheath down on them.

13

Both are sheaths inasmuch as they *cover* or protect either the knife-blade or the wings.

The name of the beetle family is Coleoptera, which word is made up from *koleos,* a sheath, and *pteron,* a wing. This combination was invented by Mr. John Ray, an English naturalist, in 1705, and it has stuck to the beetles ever since.

The larvæ or baby beetles are not caterpillars, but are generally known as grub-worms or meal-worms or wire-worms because of their worm-like appearance. Usually the larvæ have six legs near the front of the body, one pair of legs for each of the first three divisions of the body, although the grubs of some species are legless and some, one might say, *very nearly* have legs on the tail end of the body, and in many of the babies their walking, creeping or crawling is aided by warts on the belly of the grub which serve as legs and feet.

The baby beetles, like their parents, have mouth parts built for biting and most of these babies are so timid and modest that they hide themselves away from sight in rotten stumps, in the earth, under stones, inside of seeds, nuts and in acorns, in furs, woolens and hair goods. Some lead the lives of lions and tigers, catching and eating other insects, some live on land and some in the water and a few of them are degraded parasites—that is, dead-beats, insects that live on other insects, not by hunting and devouring them as do tiger beetles, but living on the bodies of other insects as do ticks, fleas, and lice upon the bodies of mammals.

Those beetles the grubs of which live in rich earth or rotten wood usually make themselves cocoons by collecting the rubbish and bits of wood around them to protect them while they lie in the mummy or pupa state, and some of the larvæ of the beetles spin cocoons much the same as do the larvæ of the moths.

In killing the beetles for your cabinet collection, the cyanide bottle does the quickest work, but it may spoil the color of the pretty red and yellow beetles. Alcohol, however, will kill the beetles and, if they are not kept in the alcohol bottle too long, it will not cause the colors to fade. Some people use a stout cloth insect net and go on a blind hunt by sweeping the grass and bushes with this net and then dumping the contents, rubbish and all, into the poison bottle, which kills the insects so that they may be removed at leisure.

A great many beetles may be collected in the springtime by scooping up the rubbish in the woods and paths, putting it in a sieve and shaking the latter over a piece of white paper.

Of course all the finer bits of rubbish will fall on the paper, but with them will come a lot of sleepy beetles which have been dozing away all winter under the leaves.

The driftwood and rubbish left by the brooks, streams and rivers on their shores may be examined in the same way for specimens. Sometimes a drop of ammonia water on a pile of rubbish, like the poison gas used by the Germans, will force the beetles to leave their hiding place and crawl on the white paper spread there for that purpose.

Many insects, including some beetles, have a habit, when frightened, of letting go all hold and dropping to the ground and thus escaping capture; but knowing this habit of theirs, the collector will often invert an umbrella—that is, put an umbrella upside down under a bush and then strike the bush with his hand and thus frighten the beetles until the foolish things drop into the trap prepared for them.

Some naturalists carry a bottle of alcohol with a cork which has a hole in it, and in this hole a tin funnel is thrust (Fig. 170). They use this novel collecting bottle for those beetles which have the

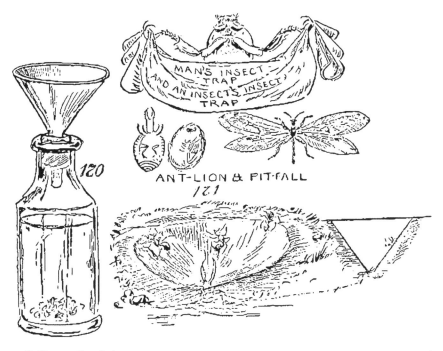

tumbling habit, especially those which infest mushrooms and toadstools. A likely mushroom or toadstool is carefully plucked, then carefully held over the top of the funnel; when all is ready, the collector fillips with his finger the toadstool, the jar frightens the beetle and the hapless insect lets go and drops, but instead of falling on the ground, it hits the tin sides of the funnel and goes slipping and sliding down into the trap through the nozzle of the funnel into the bottle of alcohol where it miserably perishes, as does an ant when it falls into the hole of a Doodle-bug (ant lion) (Fig. 171).

Water, meat-eating beetles may be collected by placing in the water a dead mole, mouse or something of that kind, or they may be seined for with pieces of wire netting as already described in the forepart of this book, and they may often be collected at night under the electric light. In fact, some of them are so attracted by the electric light that they have lately received the name of " electric-light " bugs.

But the handling of the dangerous poison bottle and the pinning of the dead beetles is not as interesting as the keeping and studying of the live ones. *There is nothing so interesting as life!* Nevertheless we need collections, in order to label and name our specimens and learn their parts, and thus fix them in our minds. In the front part of this book under " Collecting " you are told how to make a cyanide poison bottle, but I neglected to caution the reader against making the layer over the poison so thick that the expansion and contraction of the plaster of Paris may crack the bottle.

After the pieces of potassic cyanide are put in the bottle by the druggist, on top of the cyanide sprinkle the dry plaster of Paris; level the plaster by shaking it down a little, then take a common atomizer, fill it with water and spray the plaster with it. When " fixed " the plaster will hold together in the form of a shell over the poison and the shell can be regulated and should not be thicker than the glass of the bottle itself. Let the druggist do all this for you because cyanide is a dangerous poison.

When you pin your dead beetles, thrust the pin through the right elytron (Fig. 172) (wing cover) about a third of the way down and, allowing the point of the pin to come out on the right side between the middle legs and the hind legs (Fig. 173) push the beetle up the pin, leaving only enough of the latter protruding above its back to give you a hold with your fingers when you put the specimen in the cabinet or take it out.

Probably the most interesting pets in the way of beetles are the ones you find in the water. They are little trouble to feed and keep in confinement because one can put them in an aquarium (Fig.

174) where they may be observed all the time. But since the water beetles will come out at night to fly around, the aquarium should be protected by a wire netting. Some of the smaller water beetles

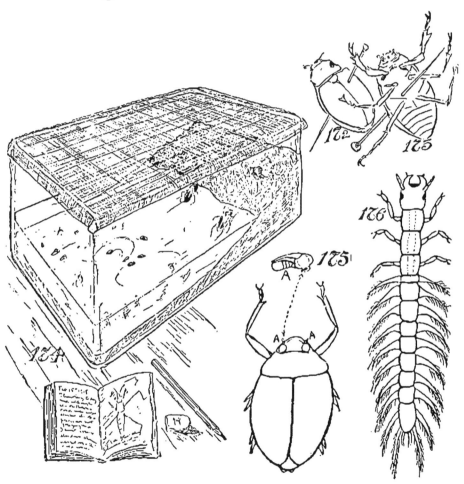

have an odd habit of swimming around and around on the top of the water in the aquarium, all the time emitting a whining, complaining noise. Others, like the whirligig beetle (Fig. 175), for instance, strenuously object to being confined in the aquarium, but will become accustomed to it in time, and so tame that they may be fed from one's hands. The whirligigs in parts of the Southwest are called " apple bugs," not because they love apples, but because when held in the closed hand for a while they emit an odor like that of sweet apples; but Packard says that when caught they give out a *disagreeable* fluid; this may be true of Yankee whirligigs but it is not true of the ones I caught as a boy on Brookshaws Pond or the Licking River in Kentucky.

The whirligig is an extremely shiny beetle of oval form (Fig. 175) and bluish-black color that you will find on the quiet eddies of the brooks, and on the surface of the ponds, where they collect in crowds composed of many individuals. If approached quietly and carefully, they will often be seen resting perfectly still upon the surface of the water, but the moment they are disturbed they start rapidly circling around in and out among themselves in a most bewildering manner.

The captives that I had in the aquarium, being unable to circle around in the wide spirals to which they were accustomed on the open water, would dive down under the water when frightened, and, clinging to a plant, remain there for some time. But after a while they became accustomed to my presence and when I caught a fly and held it for them, they would take it from my fingers, and in the winter time after the flies had disappeared they would take little bits of fresh meat from my fingers.

But the eels that lived in the sand in the bottom of the aquarium would smell the food and come wiggling to the surface of the water in search of it. The eels were extremely small, no larger than small leeches, so when they seized the food which the whirligig beetles held, it made an interesting and even fight. The eels often won, however, by twirling themselves around rapidly like a corkscrew until they threw the whirligig in the air.

The female whirligig lays her cylinder-shaped eggs on the leaves of water plants, placing them end to end in parallel lines and in a little over a week they hatch out creatures looking like thousand-legged worms (Fig. 176), each division of the body having a thread-like breathing apparatus very much like the Hellgramites, Dobsons, Clippers or Bogarts. In August these queer things creep out on the shore and spin cocoons in the retirement of which the pupa stays a month remodelling itself into the form of a beetle.

These little incidents are what give interest, they are the things that happen in life, and that is the reason I tell you boys that *live* specimens are much more interesting than dead ones. When I was a small chap like you fellows I used to make myself little cages for menageries of beetles, and sometimes used two thin disks of cork for the top and bottom of the cage and long bright pins for bars (Fig. 177).

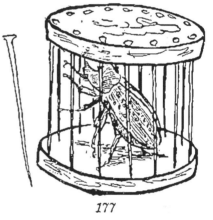

177

To-day, however, you have the wire-screen netting with which to make cages of all kinds, whereas when we boys of yesterday were building cages for wild beetles we had only mosquito netting.

An ordinary square glass aquarium, the bottom of which is covered with a layer of sand an inch and one-half thick (Fig. 174) and one end of which is banked up with sand and moss half way up the side, may be made into a land-and-water affair by putting in enough water to cover the sand and allowing the moss to serve as the land. I have such an aquarium in the window now and all winter I kept water beetles and other interesting aquatic insects with some water bugs in it.

It is my impression now that the water bugs were the victors, for along towards spring I had neglected my aquarium for some time and when I looked in it for specimens from which to make drawings for this book, the only two live creatures left were two water bugs. I do not think the other creatures died of starvation, but I strongly suspect that the water bugs sucked the juice out of them; even the caddice worms and snails were sacrificed.

The animals which prey upon other animals, as do the lions, tigers and wolves among mammals, the hawks and eagles among birds, and various beetles, bugs and spiders among the insects, are called " predaceous." Most of the predaceous insects are useful to man because they help destroy their insect relatives which live on the leaves of our trees and garden truck.

THE DIVING BEETLES

The Diving beetles (Figs. 178 and 179), the larvæ of which are called Water Tigers (Fig. 180), differ from the ground beetles in the form of the hinder sockets and shields which join the legs to the body. These are very large, touching each other on the inner edge and reaching the side of the body, entirely cutting off the belly divisions from that part called the Metathorax.*

They have oar-like swimming legs decorated with long hairs. The hind pair are flattened like a paddle or oar blade. The young are hose-shaped with big flat heads armed with pruning-knife-like jaws with which they grab their prey or even cut off the pollywogs' tails. Sometimes they catch small minnows and suck their blood.

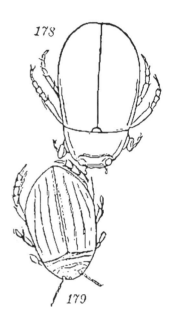

THE WATER TIGER

The body of the Water Tiger ends in a pair of long breathing tubes (Fig. 180) which it pushes up into the air. When ready for change, the larva creeps on to land, builds itself a round prison, and two or three weeks later the beetle comes out,

* That part of the chest or thorax between the upper thorax or chest and the belly or abdomen.

unless the cocoon is made in the fall, in which case they sleep all winter in it and come out as beetles in the spring.

The Water Tiger has none of the appearance of the beast from which it takes its name, but it is just as blood-thirsty. Put some in your aquariums and watch them as they go about seeking their prey and gathering air to breathe.

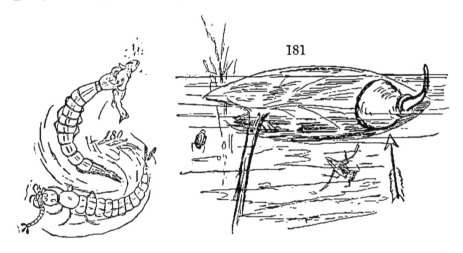

One of the most interesting facts about aquatic insects—that is, insects which live in the water—is their various ways of supplying themselves with air. Take, for instance, the tribe known as the Scavenger beetles. These beetles, when quiet at the top of the water, keep their head uppermost, as does a man. Some beetles reverse this position. The predaceous diving beetles, those whose horny

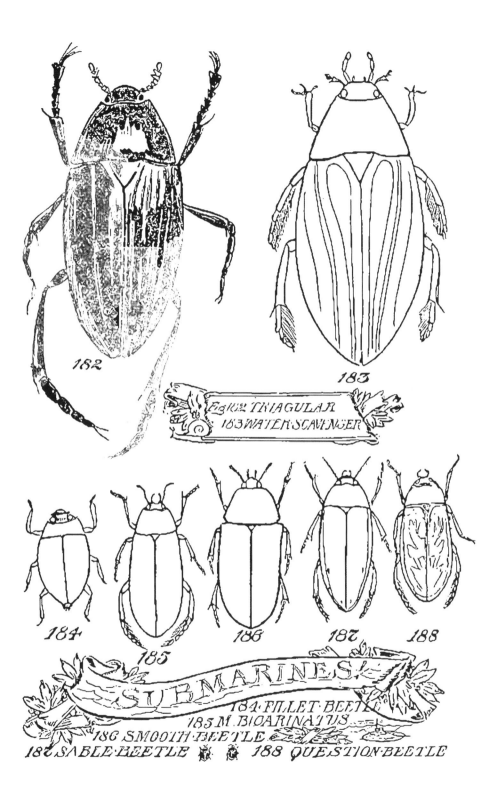

182

183

Fig 162 TRIAGULAR
183 WATER SCAVENGER

184   185   186   187   188

SUBMARINES!
184 FILLET BEETLE
185 M. BICARINATUS
186 SMOOTH BEETLE
187 SABLE BEETLE   188 QUESTION BEETLE

wing covers make a straight line where they join
on the back, rest in the water head downward, with
the tip of the tail at the surface. Many of the
insects carry the air down with them, covering the
whole under side of their bellies with minute
bubbles, which gives them the appearance of being
coated with quicksilver. When frightened, the
whirligigs hitch a bubble of air to the hind tip of
their body, and dive below with this supply of
breathing material. They remain under the water
clinging to a stone, stick or plant until more air is
needed, then come to the surface and renew the
supply.

Some water beetles deposit their eggs upon the
under side of a leaf (Fig. 181) or floating stick and
supply the eggs (and the young when hatched)
with air by enclosing the eggs in a waterproof sack
or bag in one end of which they attach a horny
pipe or tube extending up to the air.

HYDROPHILIDÆ

You can remember that name by thinking of
hydrophobia, hydroplanes and hydrants. The
Triangular (Fig. 182) is one and five-tenths inches
long and shiny black. Most of the water beetles'

14

larvæ are said to be meat eaters, but some of them, when they grow to be beetles, repent and become vegetarians. One kind is known as the Scavenger beetle (Fig. 183) because it has a very useful way of eating all the decayed matter and thus cleaning out one's aquarium. But we cannot give more space to these live submarines (yes, not only are they submarines, but also hydroplanes and aëroplanes and surface swimmers combined—Figs. 184–188), our object being only to start the reader on the road to hunting, capturing and keeping some of them alive, for besides being instructive, they are a source of endless amusement, not only to the boys who collect them but also to the parents of the boys and their guests.

CHAPTER THIRTEEN

TIGER BEETLES. HOBGOBLINS' DENS AND A REAL MAGIC TRICK. CATERPILLAR HUNTERS. BLIND HARPALUS BEETLES AND OTHER BLIND INSECTS IN MOTHER NATURE'S CAVE FOR THE BLIND. CARRION BEETLES. UNDERTAKER AND GRAVE-DIGGER BEETLES. AMUSING FACTS ABOUT CARRION BEETLES, FLIES AND ROVE BEETLES.

TIGER BEETLES

BEETLE, in old English, means a biter, and you will notice that most of the beetles can bite your finger severely enough to make you wish you had not put it against their biting apparatus. But you need not experiment with your fingers on their jaws; try beetles' "teeth" with the end of a match or broomstraw.

Among the best biters are Tiger beetles (Figs. 189-192). Every boy knows the Tiger beetle by sight, if he does not by name. Everyone has seen the lively insects running along in front of them on the sandy shore of the lake or ocean or on the dusty country road. They only run a short distance, however, then take to their wings and fly, but even then they do not go far before they alight in the road or on the beach, always facing the approaching pedestrian, and wait for him to overtake them, when they scuttle along and again take to their wings.

These beetles always attract attention because they are beautifully and strikingly decorated with metallic colors. They have large heads and large eyes and toothed jaws and they seize and feast upon the unfortunate insects which cross their path.

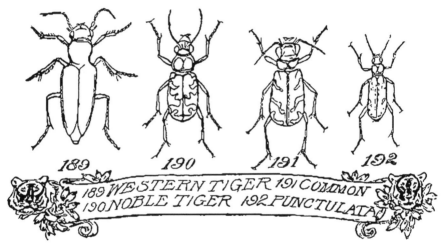

189 WESTERN TIGER   191 COMMON
190 NOBLE TIGER   192 PUNCTULATA

Even the baby Tiger beetles (Fig. 193) are meat eaters and furnished with strong jaws like their mothers and fathers. But the babies are trappers, not hunters; they lay in "watchful waiting" for their prey, dig holes in the ground (Fig. 194) creep into them and use their head for a trapdoor (Fig. 195) to cover the hole; the head being the color of the ground, it is not noticed by the

careless insect that thoughtlessly crosses the fatal ring.

I said " crosses," but it seldom gets across, it usually stops right there! (Fig. 196.) The jaws of the baby Tiger beetle which, like a spring trap (Fig. 197), have been held open, come together like a vise on the unfortunate victim's body (Fig. 196), the prisoner is then drawn into the hole and devoured at leisure.

On the fifth ring of its body, counting from the tail, the grub or baby Tiger beetle has a hump with two hooks (Fig. 193) by which the thing anchors itself in its hole when its jaws are fastened on a prey too big and strong for it to manage without an anchor, or it uses the hump to aid it in climbing to the top of its well.

If the reader will look in the paths where the ground is hard and smooth, he may find a number of small holes which have the appearance of old ant holes, but which are really holes occupied by the hobgoblin larvæ of the Tiger beetles.

My dear friend, the late W. Hamilton Gibson, once said that he counted seven small holes within sight as he sat upon the steps of his house. The reason he could count the holes was because he had frightened the hobgoblins and they had retreated to the bottom of their wells leaving the black holes in sight.

After sitting for a while on the steps, all the holes vanished, Mr. Gibson could not see one of them. The reason for this was that the hobgoblins had come to the surface and stopped up the holes with their flat dirt-colored heads, thus hiding the openings. Figs. 195 and 197 show drawings of the top of the hobgoblin's head. This head is set on the body almost at right angles, that is, with its chin down so that the head can fit like the cover to a stewpan over the opening in the ground.

You can distinguish these holes from the ordinary ant holes because each of them has a round hollow surrounding the hole, a circular trench with a central well for a retreat, in place of a hole in the ground surrounded by a hill of pellets, as have the ants. If you find some of these hobgoblins' dens you can have a lot of fun with people who know nothing about them. Point out the holes to your friends, let them count them, then make your companion sit perfectly still for five minutes or more without moving while you mutter some magic words.

Of course any words will do, but just for the sake of being accurate, you can say: "I conjure you to disappear, ye holes of the hobgoblins! Ya, Ya, Ya; He, He, He; Va, Hy, Hy; Ha, Ha, Ha; Va, Va, Va; An, An, An; Aia, Aia, Aia; El, Ay, Elebra, Elechim!" which I take from an old book of magic, so it must be right. If you do not move and keep quiet long enough the hobgoblins will come up and stop the holes with their heads, and your astonished friend will apparently see the holes disappear right before his eyes. When there are no more holes in sight, cry aloud, " I conjure the holes to reappear!" Clap your hands and stamp your foot and all the hobgoblins will disappear and leave all the holes in plain sight! This is *real magic*. It is the magic of knowing more than the other fellow.

You may fish for these hobgoblins, and when you become skilful, you can catch them by inserting a straw of grass down the hole (Fig. 194) and when the hobgoblin nips it on the end, withdraw the grass with the hobgoblin attached. In fact you can have real fun with these queer things and in doing so learn a lot about Tiger beetles.

CATERPILLAR HUNTERS

The caterpillar hunter (Fig. 198) is a long-legged beetle with powerful long hooked jaws. The caterpillar hunter is fond of canker-worms and if we had enough caterpillar hunters to eat up the canker-worms, we could save many of our fruit trees from destruction, and, if all you boys learned to know these useful beetles, you could do much

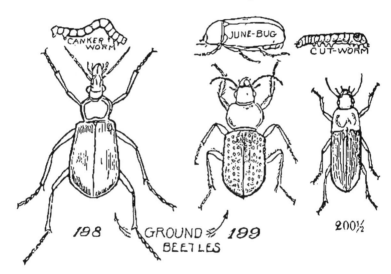

198   GROUND 199   200½
BEETLES

to prevent thoughtless people from thinking them to be harmful and killing them as bugs.

There is little danger of people killing many, if any, of the bright-colored Tiger beetles which run ahead of you on the dusty or sandy shore, because these gaudy meat-eaters are very difficult to capture even with a net, but some of the ground beetles do not fly and some of them have no wings, and they can be trampled to death as they are running along the grass in search of canker worms.

These beetles are of a dull metallic color and have a habit of prowling through the grass or hiding under sticks and stones. After dark they go hunting game. The fierce Calisoma (Fig. 199) will even attack the big June bug and rip open its sides. The June bug is a helpless brown beetle, but so big that one would not expect the other beetle to attack it.

THE HARPALUS BEETLE

There is an interesting little Harpalus beetle with a small head, a heart-shaped waist with a wide hoop-skirt effect (Fig. 200). Of course, the heart-shaped part is not the body, it is what is called the pro-thorax, but nevertheless it looks as if it might correspond with the bust and waist of a woman and the lower part represents her skirts. The little beetles are dressed in yellowish-red waists and blue or green tinged skirts—in other words, wing covers. These funny little beetles are known as Bombardiers, from the habit they have of discharging a pungent fluid with a report like a teeny, weeny gun. The shooting is probably done as a means of defense against just such enemies as the fierce Calisoma beetle might be.

Some of the Harpalus beetles do not look at all like the Bombardiers, for they are large, heavy-set individuals with an almost square pro-thorax. Every time you meet one, smooth him on the back and tell him what a fine fellow he is, because these beetles feed on cut-worms, which any man who has run a garden knows are the sort of garden sub-marines which loaf around under ground ready to attack a neutral, and the meanest and most annoying insects on a farm.

There is a funny Harpalus beetle without eyes which inhabits the Mammoth Cave in Kentucky. There are no gardens, no beds of radishes or lettuce in the cave, but for aught we know there may be blind cut-worms there for the blind Harpalus to feed upon. There are blind fish and blind craw-fish, and in some Kentucky caves I have visited I have seen thousands of blind katydids, so there is no reason why there should not be blind cut-worms and it is a pity that all of them are not deaf, dumb, blind and paralyzed.

CARRION BEETLES

Among the insects we have various trades and professions, including divers, swimmers, mud-daubers, paper-makers, net makers, scavengers, and now we come to sextons, undertakers and grave-diggers, a useful but unpleasant lot of little people. Useful because they will quickly bury a dead shrew, mouse, frog, mole, or a dead bird, and they will also do their best to bury much larger creatures which may be found dead in the field or forest and thus prevent the carrion from poisoning the air.

The female carrion beetles lay their eggs upon the dead creatures which they bury and the young beetles hatch out on the dead bodies and immediately begin to devour the carrion. The carrion beetles may be known by their very decidedly clubbed antennæ, their flattened bodies and their disagreeable odor, not to speak of their turkey-buzzard habits. The larvæ or young (Fig. 201) are long-jointed creatures reminding one very forcibly of some sort of crustacean (a family to which lobsters, crawfish and shrimp belong). The larva makes itself an oval cocoon, into which it retires while it is undergoing the change which makes it into a beetle (Fig. 202). In that asylum

Mother Nature has made for her blind creatures, known to us as the Mammoth Cave of Kentucky, the carrion beetles have sent one of their blind relatives.

The carrion beetles have a black, nasty fluid with which they are only too generous and it makes

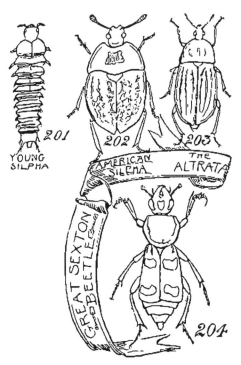

201
YOUNG
SILPHA
202 AMERICAN SILPHA
203 THE ALTRATA
GREAT SEXTON BEETLE
204

them disagreeable to handle, which is probably its purpose. Needless to say that it does not add to their attractiveness, neither does the fetid odor which emanates from their flat bodies and from their larvæ recommend them to us as pets; but in spite of their ghoulish tendencies and

offensive odor, which they retain even when dried and pinned, many of them are marked with brilliant colors, like the red-spotted Great Sexton (Fig. 204) and it is quite interesting to watch them at their work burying some small dead creature. Although I cannot recommend them as pets, nevertheless if you are making a collection of beetles it will not do to be too squeamish, besides which the carrion beetles look quite attractive in a cabinet.

We do not know positively how the carrion beetles find the dead animals, but it is supposed to be by the sense of smell. If this is true, they are much more expert than the carrion flies. If the cook is boiling cabbage, the blue-bottle flies will mistake the odor of the succulent vegetable for something much more disagreeable and offensive, and the flies will fill the kitchen with their buzzing bodies unless the screens are kept down.

Of course I do not mean literally fill the kitchen; to be more guarded in my statement it may be well to say that a great many will find their way into the kitchen to the annoyance of the housekeeper.

Out in the woods of Pike County, Pennsylvania, high in that mountainous country, I have a log house; log houses have many cracks and crevices through which small creatures may creep; when we cook cabbage in the log house, no sooner does it begin to boil and the perfume pervade the air, than the blue-bottle flies begin to appear. Although there are no flies anywhere in sight when

the cabbage is put on the stove and all the windows
and doors are carefully closed, they creep under
the door and over the sill, they work themselves in
sideways through a crack below the window sill
and soon you hear them buzzing in every corner
of the room. But never, on any occasion, has the
scent of cabbage attracted the carrion beetles.

From these amusing facts it seems that either
the carrion beetles find their food by some other
means than following their noses or that they have
a finer sense of smell than has the blue-bottle fly.
Whatever the reason is, if I find a dead frog or
mouse near my log house and with a stick push the
body to one side, it will never fail to reveal several
varieties of carrion beetles scurrying around where
the dead body lay.

ROVE BEETLES

You may recognize the Rove beetle by the fact
that it has outgrown its clothes. Its skirts are too
short, they are so short that in place of skirts they
might well be called kilts, in other words the
elytra or wing covers are very short, leaving the
naked body, belly or abdomen of the insect more
than half exposed (Fig. 205). The beetle seems
conscious of its nakedness and when it runs it raises

the end of its body and moves it as if embarrassed.

The action of the beetle in elevating its tail causes the children to fear it. The Rove beetle has stout jaws, but that is not what the children fear; they are afraid that there may be a poison sting concealed in the threatening upheld tail.

Rove beetles are found about decaying substances and their babies or the larvæ look very much like their parents (Fig. 206), that is, they are nearly as well developed, or we may put it another

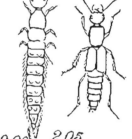

206   205

way: the parents are almost as undeveloped as the children. When the larva changes to a beetle it makes no such great change as does the whirligig beetle's larva when it changes from an aquatic worm-like creature to a round-bodied, hard-shelled, shiny beetle.

Some of the Rove beetles are as much as an inch in length, but most of them are very small. They are fond of damp places, hiding under stones, in manure heaps, among mushrooms, toadstools and moss, or under the bark and leaves of trees. Numerous species of Rove beetles dwell in ant-hills and it is possible that you may find some in the bumble bees' nest.

CHAPTER FOURTEEN

THE DESTRUCTIVE SKIN-EATERS (DERMESTES), FOND OF
ONE'S SPECIMENS, CARPETS AND FURNITURE. STAG
BEETLES OR PINCH-BUGS. THE GOLDSMITH BEETLE.
JUNE BUGS. THE SPOTTED PELIDNOTA OR GRAPE VINE
BEETLE.

BUFFALO BEETLES (DERMESTES)

THE buffalo beetles will give the collector a
lot of trouble. He will have no trouble collecting
them, for they collect themselves and will be found
to be passionately fond of a collection of other
beetles or butterflies and moths. They are oblong
and oval, with short legs, colored with white and
brick-red and black, the bottom of the elytra
(wing-covers) grayish, decorated with two broad
lines (Fig. 207).

The beetle is slow in movement, and when frightened it plays possum, that is, pretends to be dead.

It is the larvæ or grubs of this tribe which devour dried meat, skins, leather, tortoise shell and
almost any animal substance, and are exceedingly
destructive to books and furniture. Although ob-
noxious in these respects, the insects of this family
are of great service in the economy of nature, by
helping to destroy animal matter and work it into a

substance to enrich the soil and by their labors, united with those of the carrion beetles, etc., destroying such portions of these remains as are left untouched by the flesh flies that only consume the soft portions of carcasses. Like the perfect insects, their larvæ are seldom o b s e r v e d upon the surface of matters which they attack.

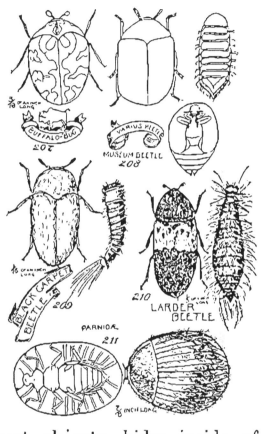

The female lays its eggs on the specimens in one's cabinet and the mean, bristly little larva eats its way into one's choicest objects, hides inside of them and eats out all the inside parts, leaving only a thin shell which falls apart with the slightest jolt. When you examine your cabinet of specimens and notice fine dust under some of them you can be sure that the baby skin-eater or dermestes is at work destroying your specimens.

Specimens which have been thoroughly touched up with poison will not be eaten by the dermestes.

Inside of some Egyptian mummies opened in 1849 were found a great number of mummy dermestes and mummy larvæ, which must have found their way there before the human mummies were prepared with preservatives.

But it isn't safe to poison your carpets on your floor and these pests will eat holes in your carpets. A mischievous dermestes has been introduced into America from Europe and we know it here as the buffalo beetle. The beetle is about three-sixteenths of an inch in length and is black, brick-red and white in color, as you will readily see if you hold a magnifying glass over one of them.

I have one of the larvæ before me as I write: It measures three-sixteenths of an inch in length. It does not seem to make much difference to the larva which way it travels. My little boy was very much amused with it, claiming that it had a head at both ends. It was caught this morning on the parlor rug, but it must have found its way there from a more secure pasture, because the parlor rug was on the clothesline being hammered by a lusty colored man only a few days ago. The presence of this little rascal, however, shows how necessary it is to keep constant watch in the summer time on all household articles made of wool.

Mr. Leland O. Howard says that the larvæ of these domestic pests are useful in destroying the eggs of the Tussock moths, also that a certain wee wasp is useful in destroying the young dermestes.

When this dermestes is outdoors it dines upon the pollen of the flowers. It is very fond of the blossoms of the shad bush. Indoors it will destroy the specimens of your cabinet and eat holes in your carpet or your clothes. It probably had more to do with introducing hard-wood floors into our buildings and doing away with carpets for our floors than any other cause. While it was plentiful fifteen years ago, it does not seem to be doing much damage at present writing. It is not fond of waxed hard-wood floors and as for rugs that people take up and shake every day, it takes no stock in them. Maybe for that reason it has again turned its attention to the outdoor world. It also has an ugly bristly larva.

BLACK CARPET BEETLE AND ITS RELATIVES

Fig. 208 shows a pest in museums, that destroys valuable specimens. Fig. 209 is the black carpet beetle, fond of feather pillows and feather

beds. It is the larva that does the mischief. Fig.
210 illustrates the larder beetle, which is very fond
of bacon and ham, and also likes dried beef. Watch
for it in May.

There is another beetle which fits in about here,
which is of an adventurous spirit. The beetle is
shaped about the same as the carpet beetle (Figs.
211, larva and beetle) and lives in the Eastern
States; the males and old-maid females are exceedingly active and when the day is hot they collect
upon the stones in mid-stream, selecting stones
that just peep out above the surface of the rush-
ing water. Here they play tag in a most lively
fashion, occasionally flying a short distance over
the water, but they do not dive beneath it. While
they frequent almost submerged objects in the
rapid water, they never allow the water to cover
them, dodging each wavelet that washes over their
particular playground. The favorite location for
them is in the dangerous waters just above Niagara
Falls. The larvæ or babies of this beetle wear a
coat of fine hair or down, which holds the air that
the babies breathe when they go below water. The
larva is shaped like a basin or shallow bowl with
an elliptical outline, that means an edge the shape

of a circle, which has been pulled out at the two
ends and made longer than a true circle or ring.
The edges of the back of this queer baby extend
far beyond the real body of the creature so as to
cover it up like a bowl. Another odd thing about
it is that it can stick its head out or draw it back at
pleasure. Yes, boys, there are a lot of funny things
in this world and this beetle is one of them.

STAG BEETLES OR PINCH-BUGS

Fig. 169 is the pinch-bug, but it is not our
native American one. Tom Sawyer never saw a
pinch-bug like that represented in 169 and we only
use it because it makes a good diagram to show the
different parts of a beetle. The male pinch-bug
has larger pinchers than the female and is a rich
mahogany color and of a truculent temper. The
fact is that this beetle knows he has a means of
defending himself; he is always armed and hence
always ready for fight. When he comes blundering into the house through an open door or a raised
screen, bangs himself against the wall and falls
on the floor, he seems to think that the wall wanted
to put up a fight of some kind, so if he is fortunate enough to fall on his feet instead of on his

back, he rears up the front end of his body, opens wide his pincers and dares everything in sight to attack him. Naturalists call him the Stag-horn beetle, but among the boys he will always remain a pinch-bug (Fig. 212). The larvæ are grub worms (Fig. 213), typical fat-tailed grub-worms with white, wrinkled, greasy-looking bodies—they

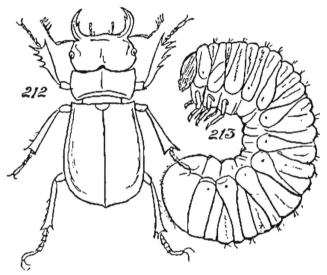

Friend of Our Youth.

look as if they would fry like salt pork. One may find them in rotten wood. When this thick white grub feels the inward call for something greater, it makes itself a cocoon of the fragments of rotten wood and retires until it comes out a real six-legged fighting stag-horned beetle, a soldier of fortune.

Speaking of soldiers reminds me of a stag-horn

of the allies of which we read in an old magazine of 1900:

"As you walk by the hedgeside a strange noise suddenly attracts your attention; it is the buzz of an insect, but loud enough to startle you; it might be mistaken for the reeling of a nightjar, but it is perhaps more like the jarring hum of a fastly-driven motor car. The reason of the noise is that the beetle has with great pains climbed up a certain height from the ground and in order to ascertain whether he has got far enough, he erects himself on his stand, lifts his wing cases, shakes out his wings, and begins to agitate them violently, turning this way and that to make sure that he has a clear space. If he then attempts to fly—it is one of his common blunders—he instantly strikes against some branch or cluster of leaves and is thrown down. The tumble does not hurt him in the least, but so greatly astonishes him that he remains motionless a good while, then recovering his senses, he begins to ascend again. At length, after a good many accidents and adventures by the way, he gets on to the topmost twig, and after some buzzing to get up steam, launches himself heavily on the air and goes away in grand style." This proves him to be a real cousin to our pinch-bug.

THE GOLDSMITH BEETLE

In the swinging drawer of my easel, amid paint brushes and crow-quill pens, lead pencils, crayons, scales, dividers and whetstones, there lies a poor crippled beetle. It is seven-eighths of an inch in length with wing covers of a light lemon yellow color and a chest of red gold with a glittering sheen, while underneath it is a metallic-green color coated with white wool. Alas! it has no legs! Something

happened to it before it was picked up in the front yard and brought into the studio. It evidently had been out all night and met with trouble. Nothing but the sockets mark the places where legs once grew; one side of its face is damaged and yet this poor cripple, armless and legless, manages to creep slowly over a piece of rough paper, or in the bottom of the drawer. Just how it moves its body I am unable to state.

The goldsmith beetle (Fig. 214) is a very pretty insect. In its baby state it is accused of injuring the roots of the strawberry vine; they also say that it injures shade trees and orchards, but personally I have never seen them plentiful enough to do any great amount of damage.

Some time in June the female deposits her eggs under the ground, laying them singly, apparently as she digs her way down. She deposits something over a dozen rather long white eggs. The young grubs come out near the middle of July.

THE JUNE-BUG OR THE MAY-BEETLE

The June-bug as the boys call it (Fig. 215), usually comes a little before June and is known among the older people as the May-beetle. The young people count it as the biggest fool in the beetle tribe, as it is always bumping and buzzing around and getting itself in trouble, banging its head against the ceiling, singeing its wings and legs over the chimney of the kerosene lamp, and apparently never doing anything with any purpose or thought.

215

These blundering beetles are of a chestnut-brown color and although the shell feels smooth to the touch, if carefully examined it will be found to be covered with little hollows, dents or dimples about the size of a needle-point. Each of the wing covers has two or more fine ridges or lines running up and down. The June-bug's breast is covered with fine long hair and the shell of the beetle seems to be thinner than that of others of its tribe.

The baby June-bug can play havoc with the clover, the hay and the lawn grass. Last season at Redding, Connecticut, they seriously injured even the pasture lands, leaving big brown patches of dead grass. Underneath the sod on the lawn one could pick up a handful of fat, white, greasy grubs in a square foot of ground. . The chickens and birds grew fat, but the farmers grew lean. The crows ate great numbers of the beetles and the skunks were not slow in hunting them. Sometimes the June-bugs injure the trees, but they are such fools, such stupid things, that if one spreads sheets under the trees in the morning, then shakes the branches, they will all fall down in a heap and may be gathered up like apples, crushed and fed to the chickens.

SPOTTED PELIDNOTA OR GRAPE-VINE BEETLE

Harris says that the grape-vine beetle (Fig. 216) sometimes proves very injurious to the vine, but the writer has never seen them in numbers suffi-

cient to do any material damage. The grape-vine
beetle has always been the plaything or playmate
of the idle schoolboy. As this beetle flies in the
daytime and is not stupid like the June-bug, it
affords more amusement. The lads tie a thread
around the body of the beetle between its arms and
its first pair of legs along the line separating the
thorax from the wing covers. If the knot is drawn
too tight it will cut the beetle into two pieces, but
if it is drawn just tight enough to keep
it from slipping off and the knot
brought round to the middle of the
back as shown on page 10, it will not
interfere with the beetle's movements
at all. And so the idle boys in Ohio
and Kentucky fasten a thread to the insect about
four feet long and the other end of the thread to a
switch or wand which they carry in their hand while
the beetle flies around overhead, to the boys' great
delight. The grape-vine beetle is a yellowish-brown
color with three dots on each wing cover and two
dots on its thorax. Underneath, the body is a
metallic or bronze green. The male is smaller than
the female and more inclined to be red, while the
female is larger than the male and more inclined to

yellow in color. The baby grape-vine beetles are
grub-worms which live in the rotten roots of trees.
You can find the beetles by looking on the under
side of the grape-vine leaves along in midsummer
and you can keep the beetles alive for an indefinite
time if you feed them with fresh grape-vine leaves.

Separated from the last-named beetle by the
fact that it has nine joints in its antenna, smeller
or feeler, and wing covers with a skinny margin or
edges, is another beetle known as

ANOMALA

This one is said to be a serious foe to the grape-
vines in some parts of our country. The larvæ eat
away the flowers, buds and blossoms of the grape-
vine. You may find them also in the sumac
blossoms.

CHAPTER FIFTEEN

TUMBLE-BUGS USEFUL AS SCAVENGERS. A NOVEL METHOD
OF MAKING MODERN ANTIQUE SCARABS. SAWHORN
BEETLES, SNAP-BUGS OR SPRING BEETLES. A SNAP-BUG
SPIRIT SÉANCE. FIRE-FLIES OR LIGHTNING BUGS.

TUMBLE-BUGS

THESE are industrious, intelligent, comical fellows and the tumble-bugs in the Ohio River Valley are a constant source of entertainment and amusement to the young people. On the steep bank of the Licking River the boys would often force the industrious beetles to roll their precious ball containing the egg (Fig. 217) which was to hatch out a baby tumble-bug (Fig. 218) over the edge of the bank and then watch the worried parent beetles hunt for the ball. If the latter did not roll too far they would find it without assistance and use every endeavor to boost it up again on the top of the bank. Sometimes they were successful, sometimes the boys had to help them.

Tumble-bugs are useful scavengers; they clean up and bury the refuse, and make their balls of cow manure, that is, the tumble-bugs of the Ohio Valley do so. One bug stands on its hands and pushes the ball with its hind legs, on the other side of the ball the other bug stands on its hind legs and pulls the ball with its hands. The ball is covered with earth, dust or sand so that there is nothing disagreeable about it any more than if it were a clay ball. The balls are buried by the beetles, sometimes many inches below the surface of the ground. The eggs hatch out inside the ball and the grub eats the material of which the ball is made (Fig. 218, larva full-grown, ready for a change). I believe there is but one egg in each ball and the grub stays in its case until it changes into a tumble-bug.

There are a number of different beetles which we might call manure beetles in the United States, some that I have seen in Alabama and Mississippi are very brilliantly colored, some have a horn like a rhinoceros. They all belong to the same family with the sacred scarabæus of Egypt, the sacred tumble-bug which is engraved on gems, sculptured in the stones and was made into necklaces and all sorts of ornaments by the ancient Egyptians. The old pottery, stone or precious-stone scarabs are considered very valuable relics and bring big prices, but it is rumored that some Yankee in Egypt is

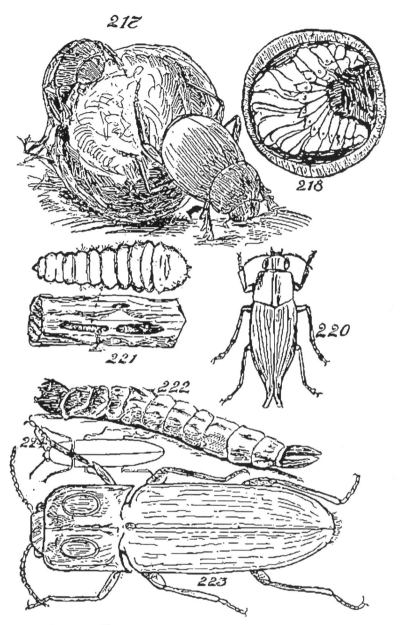

TUMBLE-"BUGS" AND YOUNG WOOD-BORER IN PINE STICK.
DICKY-BEETLE.
SNAP-BEETLES AND YOUNG.

manufacturing modern antique scarabs. It is said he has a novel method of making them look old by feeding them to turkeys, after which he sells them to the Arabs, who in turn peddle them to tourists.

SAWHORN BEETLES

These beetles form a great tribe sometimes called Serricorn beetles, but sawhorn is easier to remember. They are so called because the tips of the joints of their antennæ are thought to look like the teeth of a saw. Among the sawhorn beetles are the Dicky-bugs (Fig. 220) which the French call the Richards and some English call burn-cows and others call Bubrestians, but the Dicky-bug is the name by which the boys used to call them and it is a name one can remember, besides which Dick is short for the Richard of the French.

You will note in Fig. 169 that there is a little piece of shell shaped like a triangle up near the waist of the beetle where the wing-covers join. It is quite distinct in Fig. 169 and in most of the beetles already described, but when you come to Dicky-bugs, it is very small. The Dicky-bugs are often very prettily colored and you can find them on the branches of the trees, where they move very slowly, and if you frighten them they play possum, folding up their legs, letting go all hold and falling to the ground. But when they want to fly, they are experts at it. The young of the Dickies are sawdust eaters; they bore into the log or tree trunk, chew up the wood and swallow the sawdust. Fig. 221 shows the larvæ of some wood-boring beetles that I found eating a dry pine stick which I was whittling. Very dry food, one would think, but the little grub seemed to grow fat on it.

The hickory borer is of a dull brassy color, but a bright copper underneath and it is thickly engraved with numerous lines, besides which it has some black spots which stick up on its wing covers and the ends of the wings separate into two points. The Dicky-bugs or beetles, as they would be properly called, damage wood of different trees. One is the Hickory Dick and then there is the Big Pine Dick; all of the tribes are injurious and do a lot of damage. They bore into the pine logs of which my log house is built. Then comes the Ichneumon fly, with a very long ovapositor (egg putter) which she pokes down into the worm hole in the log and shoots her eggs into the body of the soft grub; the little Ichneumon babies, when they hatch, eat the grub up.

I once saw an Ichneumon work over half an hour trying to put its ovapositor through the head of a nail; evidently the black spot made by the head of the nail was mistaken for a worm hole by the Ichneumon.

SNAP-BUGS (SPRING BEETLES)

It is too bad that the name " bug " should be attached to all these beetles; we know they are not bugs and snap-beetle would sound just as well as snap-bug. But bugs is what they are called, and we must follow suit even if we know better. The finest of all snap-bugs is the big gray one with the eye-spots on his shoulder blades (Fig. 222 larva, Fig. 223 beetle).

Sometimes the snap-beetle is called the Death Watch (Fig. 224) and when superstitious and ignorant people hear the snapping on the walls of an old house, they are sure that means someone is going to die soon, someone who is living in that house is going to die! If you told one of these people that it was only a snap-bug calling its mate it would do no good; they have been taught that it is the Death Watch and they will believe it is the Death Watch as long as they live.

If this foolish belief would end there, we would not care, but these people will try to teach you boys that the noise of the snap-beetle is the Death Watch and it is one of the purposes of this book to set you right on this question and many others like it.

When the old red-headed woodpecker hammers on its drum, the hollow tree, or the yellow hammer, highholder or flicker does the same, it is calling its mate. The rat-tat-tat has the same meaning to the woodpecker or yellow hammer as did the plinkty-plunk of the troubadour's lute to his fair lady. And that is all the meaning that there is in the snapping of the beetle.

But if you want to have some fun with a snapping beetle, get one of the smaller kind, one of those little brown fellows or the ash-colored snap-bug (ash-colored Elater). Keep him in a little pill box or some convenient place until evening, then when the family is looking for amusement, tell them all that you are a medium and the spirits will rap for you on the table. Have the company sit around the table and only rest the tips of their fingers upon it so that there will be no cheating (?). Under your finger you have Mr. Snap-bug (Fig. 224), back down; a slight pressure will cause him to make a decided rapping noise. In all well-regulated spiritual séances they began by saying: "If there are any spirits present they will please manifest themselves by rapping." This is the time for your snap-bug to answer. Then you ask the spirit to rap once for Yes and twice for No, after which you can ask any question you choose and get just the answer you want, at the same time greatly astonishing and mystifying the rest of the circle. I am telling you this to show you that even a lowly snap-bug, a wood borer, an outlaw, is of some use in the world, for anything which can serve the purpose of harmless amusement is doing the world a great service. After the snap-beetles come the fire-flies and these fire-flies are no more flies than are the snap-bugs bugs; they are all of them beetles.

FIRE-FLIES

Of course the fire-flies, like all other creatures, have a lot of relatives; they really are, I believe, only a sub-family, but the lamps they carry give them a distinction which their relatives cannot claim. The light-giving organs are not always in the same place on different kinds of fire-flies, or as we always knew them in the middle West, lightning bugs. The babies or larvæ as well as the beetles are luminous and some people say that the eggs give light, but this doubtful. If you mash a lightning bug the light is brighter than before you stepped on it. The Pennsylvania lightning bug is about five-tenths of an inch long, and of a sort of

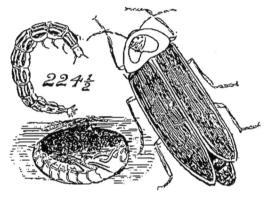

yellowish color with dark colored stripes. It is the lightning bugs (Fig. 224½) which lend such charm and enchantment to the field and roadsides on summer nights. The little fireworks people are soft-winged beetles of the family Lampyridæ, which have the property, the gift, or the power of sending out from their bodies flashes of soft light. There are several distinct species of so-called fire-flies native to North America, according to the eminent naturalist, Professor Riley, the most common and widely distributed of which is Photinus pyralis (Linn.). This insect is most abundant in the Southwest, where, during summer evenings its constant flashes of light give the air the appearance of being filled with moving sparks of fire. The beetle is of oblong form, somewhat flattened and varies from one-half to five-eighths of an inch in length. It has a dull black wing covered with pale yellow edges, a yellow chest with a central black spot set in patches of rose color. The under side of the abdomen is dark brown with the exception of the two end rings, from which the light is sent out; these are sulphur yellow.

If you live in the southwest middle states, note the way the lightning bugs give out their light while on the wing, then when you travel into Yankee land note the way the lightning bugs there send out their light and the way they do it down in Mississippi. Some of them emit light as they make a downward dash, thus making a streak of lightning, suggesting the name of lightning bug, while others seem to glimmer, glow, increase gradually in intensity of light—the light grows brighter and then gradually fades out again.

You should, if possible, collect the fire-flies from all these different sections of the country, and this you can do by trading with boys who live in other sections; you will find that the beetles differ in their markings and some other respects, as well as in their actions. In Kentucky I have seen the little girls, after dark, wearing organdie, tarlatan and lawn dresses between which and their skirts they had inserted a number of lightning bugs, producing a very pretty effect as the insects flashed their signals. I may be wrong with regard to the name of the cloth the girls wore—I am not an expert in dry-goods—but it was a thin, flimsy material and showed the light emitted by the insects almost as plainly as if there were no cover over them. The lightning bug furnishes safe and sane Fourth of July fireworks.

CHAPTER SIXTEEN

DEAD-BEAT STYLOPS. WEEVILS. PEA WEEVILS AND OTHER
EVILS. BALTIMORE ORIOLE'S FONDNESS FOR GRUB OF
THE PEA WEEVIL. GOAT- OR CAPRICORN-BEETLES. LEAF-
BEETLES. POTATO-BUGS. ELM-BEETLES. UNDESIRABLE
CITIZENS AND LADY-BUGS

DEAD-BEAT STYLOPS

THE Stylops is a warning to all such people as
have a desire to live on someone else, to sponge on
someone else for a living in place of paddling their
own canoe. The Stylops is a degraded dead-beat
and a criminal among insects. Take a look at
Fig. 225 and ask yourself how you would like to
be Stylopized. I want you particularly to look at
the intelligent (?) graceful (?) and fascinating (?)
form of the female Stylops (Fig. 226). You see
she does not need brains, she does not need feet
or antennæ, she needs nothing but a digestive tube
because she lives inside the body of a bee (Fig.
227) and the bee has to do all the work and all
the fighting.

When the Stylops is young it has legs and
can run, but it chooses the life of a dead-beat
and the dead-beat life has degraded it.

The young are hatched inside the body of the

251

parent and the parent is therefore called viviparous. She gives birth to about three hundred babies at a time, not counting those which get away unobserved. Inasmuch as neither the old lady nor the old gentleman Stylops have to support their children, they can afford to have big families.

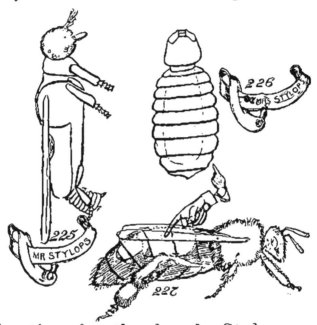

In hunting for the female Stylops you must examine the bodies of bees, where you will sometimes find the head of the fat criminal sticking out from between the abdomen plates or the belly rings of the bee.

The male Stylops looks like Mr. Pinheadus in the comic sheets of the newspaper and he is a pinhead among the beetles. He has wings and an

excuse for wing-covers. Mr. Pinheadus dresses in
a black suit with a short brownish-colored vest.
He measures about one-fourth of an inch in length
and much less in intellect.

WEEVILS

As a rule these beetles are very small, but with
few exceptions have exceedingly long noses (Fig.

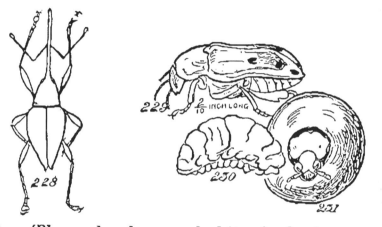

228). They also have a habit of playing possum
like some of the beetles already described. There
is a pea weevil (Fig. 229) which lays its eggs on
the pea blossoms and the grub (Fig. 230) eats our
green peas. It stays in the seeds of the pea all
winter and comes out the next spring as a weevil
(Fig. 231) unless the summer is hot and dry, in
which case it may come out in the autumn. This
beetle is a short-nosed one and is about one-fifth

of an inch in length with a rusty black color mixed with more or less white on the wing covers. A side view of the insect is shown many times enlarged by Fig. 229. This weevil was first observed near Philadelphia, from which place it spread to most of the States where peas are grown. When the peas are in bloom the beetle appears, and while the pods are growing rapidly the females deposit their eggs upon any part of the surface, making no attempt to insert them within the young peas. The eggs are of a yellow color and fastened to the pod by means of a mucilage that the weevil supplies, which when it dries has the lustre of silk. " Pods will often be found to have from ten to twenty such eggs deposited upon them and later the young larvæ may be seen through the thin transparent shells." The larva soon makes its way through the pod into the nearest pea, the place of its entrance being a small spot, like a pin hole. The larva feeds upon the pea but avoids the germ and, with a wonderful knowledge of its future needs, eats a circular hole on one side of the pea, leaving only the hull as the covering, or ready-made cocoon. After this it passes into the mummy or pupa state and at last becomes a beetle. When ready to come out the mature insect needs only to cut the thin husk and it is free.

Up in the elm tree there is a swinging nest. The head of the family, the gentleman, is colored orange and black, the colors of Lord Baltimore, and the bird is known as the Baltimore oriole, which is very unfair to the bird because he had those colors thousands of years before Lord Baltimore or his tribe were born. But, be that as it may, the Baltimore oriole is familiar with the habits of the pea weevil and will split open the pea pods and eat the grub. Ignorant people think that the oriole is an enemy to the peas and that he splits open the pods to eat the seeds.

There is a rice weevil, which feeds on rice, wheat, and even corn, and a plum weevil, a white-pine weevil and a long-snouted nut weevil. There seems to be a weevil for everything and maybe it would not be far amiss if, in place of weevils, we called them evils, long-nosed evils. No doubt there is a reason for their being on earth, but that reason is not for the good or protection of our gardens. I doubt if the weevil is of any service to the farmer, but there is little or no doubt that the farmer is of great service to the weevil.

GOAT-BEETLES—CAPRICORN-BEETLES

These beetles will not butt you and they will not make a noise like a nanny goat. They are called Capricorn- or goat-beetle because their antennæ are long and often bend back in a curve like the horns of a goat (Fig. 232). The bodies of the goat-beetles are generally long and rounded. Their short heads are armed with powerful jaws. I have

232

seen one of them grasp the point of a six-H lead pencil with his jaws and hold the pencil erect, a feat of strength which would make Samson's work child's play by comparison. They might be called the beetles with the iron jaws.

Most of the goat-beetles have long legs and four-jointed feet with wide-cushioned soles. When you pick one of them up, it will squeak like a little mouse, but insects' voices do not come from their lungs; they make a noise by rubbing some of their joints together. Goat-beetles rub their thorax and belly-joints together to make the squeaking noise.

The female Capricorn-beetles have a jointed ovapositor—that is, a jointed egg-layer in the end of their bodies which works like the joints of a telescope. When they want to put eggs into any crevice, crack or hole in the wood or plant, they run out their telescope, insert it in the hole and then shoot their eggs into the place where they wish them to be.

The babies are long grubs, whitish and fleshy with the rings of the body very convex—that is, arched-like or as Harris says "hunched up both above and below."

Although these babies have a small head, it is provided with short but very powerful jaws, so powerful that it can tunnel its way through the best of solid wood. These borers will make holes in the logs of your cabin, especially the bottom logs where the dampness comes up from the earth. Some of them fill the hole up behind them with castings known by the name of powder post and many of them live for several years in the log before coming out as beetles. Others of the borers keep the back door open and below it you will find a little pyramid of fine sawdust.

There are several families of Capricorn-beetles,

but we will consider them as one family to save
time and space. One of the biggest of the goat-
beetles is the broad-necked Prionus (Fig. 233 beetle,
Fig. 234 pupa), a long coal-black fellow with
thick and stout jaws and thick and saw-toothed
antennæ. The goat-beetles choose different trees
in which to make their gimlet hole.

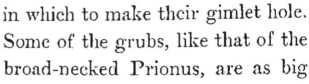

Some of the grubs, like that of the
broad-necked Prionus, are as big

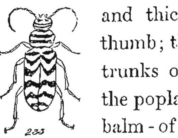

and thick as a man's
thumb; these live in the
trunks or the roots of
the poplar trees and the
balm - of - Gilead trees.
Fig. 235 is the common
golden-rod beetle.

Like the weevil, they
seem to adapt themselves to all different trees,
being loath to slight any. One of the largest goat-
beetles found in New England is the tickler (Figs.
236 and 237), so named on account of the habit
which he has of waving his long antennæ and gently
touching with their tips the surface on which the
beetle walks. When they are courting, they wave
their long antennæ around in a graceful manner

and make a creaking noise. Fig. 238 is the pretty blue-and-yellow elder beetle.

The goat-beetles seem to be often afflicted with what the doctors call arrested development. That is, their babies stay babies for a long time. Away back in 1889 it was reported that the State Ento-

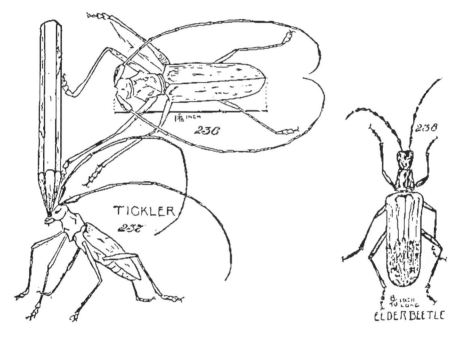

TICKLER
236
237
238
ELDER BEETLE

mologist of New York had sent to him a beetle which had bored holes through a kitchen painted floor at a place called Howe's Cave. The holes in the floor were about a quarter of an inch in diameter. The beetle itself is the long gray fellow with black dashes on its wing covers known as the Long-horn pine borer. The baby larva or grub of this pine borer is the one that ruins so much good pine lumber. In the present case the grubs must have been in the pine logs before they went through the saw-mill and were made into flooring boards.

The grub must have taken some Rip Van Winkle " naps " which made it sleep and of course remain a baby for a long time. In the Peabody Academy of Science at Salem, Mass., one of these beetles is preserved which had eaten its way out of a blue bureau which was made fifteen years before. As showing a greater imprisonment in furniture, it is traditionally said that in 1786 a son of Gen. Isaac Putnam, residing in Williams-town, Mass., had a table made from one of his apple trees. Out of this table, twenty years afterward, a long-horned beetle gnawed its way, and a second one burrowed his way out twenty-eight years after the tree was cut down.

LEAF BEETLES (CHRYSOMELIDÆ)

The leaf beetles are longer than they are wide; egg-shaped, sometimes are very thick through the body, the back is rounded like the half of an egg which has been split endways, the eyes are promi-nent, their chests are narrow and cylindrical. The upper part of the hind legs are sometimes divided in the middle, and the belly has five free rings.

The babies are short, sometimes cylindrical, and sometimes flattened, often brightly colored, usually soft and mushy and ornamented with flattened warts or branching spines. I am giving you these general items because it is calculated that there are between eight and ten thousand species and we can have but a few drawings. The leaf beetles are feeders on the leaves of plants both when they are insects in the perfect form and in the larva state.

ELM BEETLE

Of course, every boy knows some one of the elm beetles, the larva of which strips all the leaves from our elm trees, then, while the poor trees are gathering strength to put out a new crop of leaves, the elm beetles are getting ready a new crop of baby beetles to eat up the leaves as soon as they appear, and the rascals keep up such tactics until they eventually kill the goose that lays the golden egg; in other words, the elm tree which furnishes them their food (Fig. 239, larva; Fig. 240, pupa; Fig. 241, beetle).

There are a number of so-called elm beetles,

one of which is seen in Fig. 241. This fellow is a
yellowish brown with indistinct dark stripes. All
the elm beetles are undesirable citizens and should
have been sent back to Europe when they arrived
at Ellis Island, if that is the place where they did
arrive. To tell the truth, they probably slipped in
at some unguarded port and did not come with
the regular line of immigrants.

But if the newspaper reports can be relied upon,
and I doubt it in this case, the New England elm
beetles are a military lot, who in 1895 came march-
ing into New Haven and also into Chester where
the people one morning met an army coming
through the principal streets of the hamlet. The
report says: "An animated dark ribbon, or the
folds of an immense serpent, billowed on past, in
tiny undulations. It was a wondrous, giant cara-
van of strange worms, belting an entire township,
which, having filled themselves with the produce
of a district further up the valley, were migrating
to a new field and pastures green. Many people
of Chester came into the street and gazed help-
lessly and with much concern at the orderly dis-
ciplined column rolling along the street at a speed
of 400 or 500 feet an hour. The worms (larvæ)

were banked densely in their narrow path and were massed two or three deep in some places while they marched twenty or thirty abreast. They wore gray-white bodies with coal-black stripes down the back, they had black heads and were three-eighths

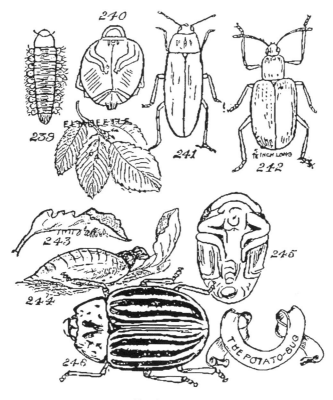

Leaf-eaters.

of an inch in length. It took them all the forenoon to go through Chester."

This description sounds like an account of a hike of army worms. Evidently there was something doing in Chester, but personally I never saw elm beetles doing anything like a Fourth of July parade or an orderly march. I have seen millions of them and seen trees stripped by them, but I never saw them move from one place to another in an army.

THE GRAPE-VINE FIDIA

This is a very prominent beetle in Missouri; it is chestnut brown in color; that is, its body is a chestnut brown but its hair is white and it is all clothed with short hair. The grape-vine Fidia has decided ideas on grape-vine leaves as an article of food, and although it will sometimes eat the leaves of the wild grape, it will not if it can help it feed on any other vine than those known as Norton's Virginia grapes and Concords (Fig. 242).

There is the asparagus-leaf beetle who is a foreigner and the apple-tree leaf beetle and the yellow hemlock beetle, and numerous others which you will find when you take up the study of beetles. There are also cucumber and squash beetles which you should know by sight.

THE COLORADO POTATO-BUG

which, of course, is a beetle and not a bug, is another undesirable citizen, but in this case it is a native American which was an Aborigine like the Indians and lived in the mountains of Colorado. It attracted little attention at first and no one knew how important it was destined to become in this world. Very few people noticed these beetles as the insects sat on the wild plant known to scientists as the Solanum rostratum which is, I believe, a plant related to our potato. Fig. 243 shows eggs attached to leaf; Fig. 244, larva or young; Fig. 245, pupa or mummy, and Fig. 246, the perfect beetle.

One day, Mr. Potato-bug woke up. Civilization and cultivated fields had reached his mountain home. This was his great opportunity and in place of a few scattered wild Solanum plants, there were scattered acres of luscious potato plants! The potato beetles literally waded into the garden plants.

Prosperity had found the potato-bugs and they followed it up until at length they reached the Atlantic Coast, where I have seen windrows of potato beetles washed up on the beach. These last were the adventurous insects who wanted to go still further east and were drowned in the attempt.

It was away back before my readers were born.

somewhere around 1855 or 1859, that the potato-bugs began to attract attention by attacking the neighboring fields and working eastward. But they took it leisurely and were in no hurry, because they were living on the fat of the land. They had nineteen years of riotous living before they reached the Atlantic Coast. The beetles have crossed the Atlantic a number of times, but they were recognized over there as undesirable citizens before they could multiply or spread.

The Colorado beetle is a striped fellow, considerably larger than a green pea, which is almost equivalent to saying as big as a piece of chalk. It is a trifle over a half an inch in length, it is almost oval and of a yellow color with black stripes and blotches. Its wings are red and show when it flies. Red is the sign of danger, of revolution, of energy, and I think this insect stands for all three (Fig. 246). Of course it is the larvæ which eats the most, but in this case the beetle also feeds upon the potato plants.

LADY-BUGS, LADY-BIRDS

They formerly used lady-bugs to cure the toothache, now they use them to cure the San José scale. This is a beetle of course and is neither a bird nor a bug, nevertheless, as children, we always said to one of the captured insects:

Lady-bug, lady bug,
Fly away home,
Your house is on fire,
And your children all gone.

It was always a bug to the American children and a bird in the Sunday-school books and with the European children, but it is a beetle to naturalists. Some time ago, an eminent botanist brought several tiny Oriental lady-bugs from China, but though he took the best of care of them, many insects died en route. Even after landing more of them perished, so that finally only two little lady-bugs remained to face the great feast of juicy scale insects.

These two, however, were carefully nourished and trained by the Government and now quite a numerous progeny is ready to take a stand against our natural enemy, the scale. The Government in using lady-bugs for this purpose is following the method of extermination used in China.

In 1888, Albert Koebele, a collector for Professor Riley, discovered in Australia a little lady-

bug of the usual reddish brown color (Fig. 247),
which greatly loved to eat scale insects. It seemed
only to care for the fluted scale (Fig. 248).
Mr. Koebele collected a great number of these lady-
bugs and a little of their food, both of which he
packed away on ice in the steamer at Sydney,
Australia. The lady-bugs reached Los Angeles,

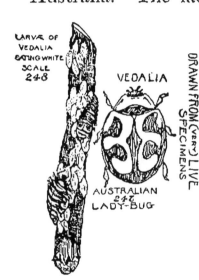

California, alive and terribly
hungry after their long trip.
They were let loose on the
scale insects there which
pestered the trees, and they instantly began to eat up these
mischievous pests, one after
another in rapid succession.
Then they began to lay eggs
and if half of the young ones
grew up to be female beetles one lady-bird would,
in six months, have 75,000,000,000 children, each
of them hungry for scale insects!

So you see, lady-bugs are of some use in the
world; even the foreign ones like those from New
Zealand do not make undesirable citizens of our
republic.

Never kill a lady-bug, a lady-bird or a lady-beetle and remember that the gentlemen beetles in this case are always known as lady-bugs too. They are probably suffragettes, and if they are they are militants. Among the scientists they are known as one-spotted lady-bugs, two-spotted lady-bugs, or nine-spotted lady-bugs, but of course scientists do not call them bugs; they have scientific names suggested by the number of spots on the beetle's back.

The lady-bugs always appear to be gentle little

creatures but that is because we are so big they do not attack us and because we do not watch them closely enough to see how fierce they are among plant lice. There is one dusky little lady-bug known as the Lion Whelp because he is so fierce. But it is fierce and bloody-thirsty only among plant lice. So the more we have of these beetles, the better it is for our rose bushes. Fig. 249 is the common Maculata; Fig. 250, larva; 251, pupa; 252, perfect Convergens beetle.

CHAPTER SEVENTEEN

BUGS, BEGINNING WITH SOME OF THE LOWEST, MOST DE-
GRADED OF THE BUG FAMILY. PARASITE DEAD-BEATS
AND OUTCAST BUGS. PLANT LICE. SCALES AND APHIDES.

BUGS (HEMIPTERA)

THERE is a very big family of insects properly
called " bugs." They are of all kinds, shapes and
sizes, some of them so different from others that
they do not appear to be relatives. But there are
certain family characteristics; for instance, their
mouths are different from the mouths of the other
insects, and to make them different, the head and
breast is altered to suit the necessity of hitching
on a horny, jointed, hollow beak to the front of the
head. This sucking tube is long, slender, and
tapering when it has to reach far into the substance
from which the bug feeds in order to get at the
juices, or it may be short and stout, according to
the food upon which the creature is dependent.

Another difference is in its wings. As a rule,
the upper half of the wing is horny and thick and
the lower half thin and skinny, more like thin
transparent tracing paper. But with bugs as with

270

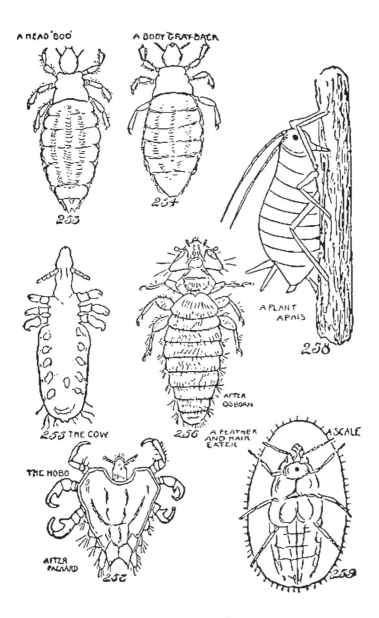

A HEAD 'BOO'

A BODY 'GRAY-BACK'

263

264

A PLANT APHIS

258

255 THE COW

256 A FEATHER
AND HAIR
EATER
AFTER
OSBORN

THE HOBO

A SCALE

AFTER
PACKARD

257

259

DEGENERATES.

other creatures they become degraded when they become dead-beats and one consequence is the parasites lose their power of flight and lose their wings altogether, hence those bugs which infest the beds in unclean houses, and infest the bodies of unclean people, very fortunately for clean people have no wings.

Among the bugs that live in the water also are some without wings and some with half-wings, and others with well-developed wings that are good fliers.

We will take the lowest and most degraded of bugs first, in order to get over the disagreeable part as quickly as possible. I suppose it is needless for me to say that I did not make the sketches (Figs. 253–256) from live specimens, for I never made a personal study of these disagreeable insects, but they belong with bugs and must find a place here, besides which I am talking to boys, and every boy is liable some time in his life to see one or more live specimens.

During the war of the States, the Union and Confederate armies were infested with these things which they called " gray-backs "; from the general and his staff down to the private in the ranks,

18

all sooner or later had some experience with them.

But in times of peace in the United States, only hoboes, tramps and those unfortunate people who have to sleep in lodging houses and cannot change their clothes often, are afflicted with these parasites. Permanent camps, like lumber camps, usually are also supplied with them, and so are the wigwams and huts of savages.

The savages' tepees, however, are not the only places one must avoid. I remember one time visiting some fishermen's shanties on the coast of Maine. My companion on this trip is now a celebrated writer and president of a famous club. He was always a neat, dapper and well-dressed man; even in those early days when he was writing stories for the newspapers he was noted as a well-dressed young man. I had my suspicions of the fishermen's cabins, and when we entered one I declined the proffered seat, but my companion, being a genial gentleman and democratic, sat down on the edge of the bunk in one of those cabins while he took notes of the yarns the fishermen told him.

When we returned in our sailboat to the rocky coast where our cottages were located, I imagined that my legs felt uncomfortable, so I waded into the ocean where it was shallow, rolled down my long woollen stockings, took off my sailor's slippers, rolled a stone on the stockings and shoes to keep them from being washed away and walked barefooted in my knickerbockers to my cottage. There was no cause for me to remember the incident but the serious yet comical consequence to my comrade.

I did not again see my friend while I was on the coast of Maine. The word was passed around that he was sick and would not see anyone, and it was not until I met him in New York that I learned why he had denied himself to all visitors. While sitting on the bunk in the fishermen's shanty the poor fellow's clothing had become alive with,—well with Fig. 254, and as he had never had any experience in this line before he did not know the cause of his trouble until he was covered with an army of Fig. 254's, and when he discovered them he was ashamed to tell anyone of his plight. He had a most serious time ridding himself of these pests, for they got into his trunk, his bed and the furniture of his room before he discovered them.

Another gentleman I knew, a dignified, wealthy New York manufacturer, had the same misfortune happen to him while sleeping on a public sofa in a public cabin when all the staterooms were occupied on the steamer. He also brought the creatures home with him and spread them broadcast before he discovered them.

I tell you these incidents as a warning to you boys so that you will avoid any similar adventures. You are liable to pick up Fig. 253 in school or almost anywhere else, because this is the one that loves to get into your hair. One careless boy can spread Fig. 253 all through a school before it is discovered, and one nurse girl can supply all the children in the family with them.

Fig. 255 is found on cows, Fig. 257 on low, degraded people, and Fig. 256 on birds. I have never had any experience with the sort that infest cows or people, but in my investigation of birds I have had my hands and arms covered with the flat white creeping things which torment our songsters with the pricking of their feet and by feeding upon their feathers. However, these insects are not built to stay on a human being, and may be brushed off, or one can rid oneself of them by a change of clothes. They are fond of birds, not people.

On account of the mouth being built for biting in place of for sucking, like the other bugs, this

Fig. 256 does not really belong with the others preceding it, but should be used as the link connecting the bugs with the grasshoppers. However, since the habits and general degraded appearance of Fig. 256 correspond with the other degenerate bugs, we place him in their company as that is the place the boys would naturally expect to find him.

A look at these diagrams is sufficient to show to what low depth even a bug can fall by becoming a parasite. As there are many bugs that are cleanly and interesting, we will leave these degenerates with the hope that our readers will never have occasion to see them anywhere but in pictures.

Their very name is not mentioned in polite circles, for all agree with Robert Burns that

"* * * * a ———, Sir, is still a —
Though it crawls on the curls of a queen."

SCALE INSECTS

For some good reason, while it is considered bad form to call by name the insects which infest slovenly beings, we can, without breaking the rules of propriety, use the same name when it is applied to plants. Thus we can speak of a plant-louse (Fig. 258) or of an oyster-shell bark-louse and not even shock the sensibilities of the most particular persons. But the mention of an oyster-shell bark-louse will sometimes cause a fruit grower and farmer to use words outlawed in well-regulated society. The reason of this is, because this scale does great damage to the fruit trees.

The scale (Fig. 259) is another undesirable citizen which emigrated from Europe to the United States. It does the most damage north of the Mason and Dixon line, and is called the oyster shell because the little thing, which is only one-twelfth of an inch in length, is something the color and very much the shape of a tiny oyster.

You will sometimes find scale insects on a potted plant in a conservatory, often on the maple and fruit trees in your yard or orchard, and they are plentiful in the green-houses of the florists, where they may be seen plastered on the bark of the orange and lemon trees. The scale is a sort of bowl-shaped shell which fits over the insect and protects it from weather and bug-eating bugs. (Fig. 259 shows the under side of one of these scales.)

Some of the scale insects are very useful. The Lacca of India produces the stuff called lac, of which sealing wax and varnishes are made. In Malabar, Bengal, and in Siam, there is a teeny-weeny mite of a scale from which the beautiful color used by artists, and known as carmine lake, is derived.

The white cotton scale often infests the branches of the soft maple, sometimes spreading from them to the grape-vine, as it did one season, to the grape-vine in our own back yard.

Another useful plant louse is the Cochineal bug, which was originally a native of Mexico and was imported from there to Spain and Algiers. We, the boys of yesterday, used to buy the dried Cochineal bugs at the drug store with which to color eggs on Easter Sunday.

The common rose-bug or Aphis is well known to all my readers who have paid any attention to the cultivation of roses. The Baltimore oriole, scarlet tanager and vireo are very fond of these plant lice and I have watched them by the hour, going carefully over a plant and picking off the little green or black bugs which were sucking the juice out of the garden flowering shrubs. The Aphis has a couple of tubes sticking out of its back, through which it can, whenever it feels like it, squeeze out a sweet substance called "honey-dew" of which the ants seem to be particularly fond.

CHAPTER EIGHTEEN

LEAF AND TREE HOPPERS. GROTESQUE AND COMICAL
INSECTS. CUCKOO-SPIT. HARVEST FLY, LOCUST AND
SEVENTEEN-YEAR LOCUSTS. A METHUSELAH AMONG
INSECTS. SEVENTEEN-YEAR LOCUSTS ATTEND A BALL
IN KENTUCKY. HOW THEY SAW HOLES IN THE TWIGS.
HOW THEY ARE PREYED UPON BY DRAGON-FLIES AND
WASPS. HARMLESS PLAYMATES. PUPA SKINS AS TOYS.

TREE-HOPPERS AND LEAF-HOPPERS

THESE insects (Figs. 260–264) have apparently
used up all their ingenuity in designing queer
fashions and forms. They indeed are an odd looking tribe, and still more weird forms live in other
countries. They feed on the sap of trees and plants
and they never know when they have enough, at
least some people claim that these insects suck up
so much of the juices that the sap oozes out of their
bodies, often hiding them in a mass of lather or
foam. In England they are called frog-hoppers,
and on account of the foamy material are sometimes known over there as cuckoo-spit, a real pretty
name (?), but I prefer leaf-hopper, don't you?

I am not prepared to say of what this foam is
composed, or whether it is really sap of the tree

leaking through the crevices of the insect's body
or whether it is something which the insect itself
produces for the sake of concealment, but I agree
with everybody else when they claim that the leaf-
hoppers are the funniest things to be found among
the insect tribes. The leaf-hoppers or tree-hoppers

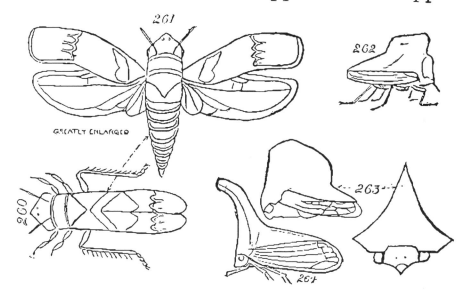

Funny Hoppers.

are the sort of bugs which could appropriately inhabit a " bug-house " for they are certainly a crazy
looking lot (Fig. 260–264).

CICADA, HARVEST-FLY, " LOCUST "

Here we are again, up against a common and
almost universal name for this well-known insect,
to which it has no right at all because the locusts,

you know, are those creatures the boys call grass-hoppers and they are not even distantly related to the Cicada, which is a bug pure and simple. Look at him and you will see the long beak under-neath his body which marks his race. But, how-ever troublesome the locust may be, there is noth-ing uncanny nor disgusting about him.

The locust is one of the most interesting of bugs, a good play-fellow and it cannot hurt you; you may play with it all you choose without offend-ing it, for it will often sing for you while you have it between your fingers.

I said that it cannot hurt you and I have good reasons for supposing that you cannot hurt it, be-cause seventeen-year locusts have been discovered blithely singing away entirely unconscious of the fact that some other insects had eaten up most of the singer's body.

It is probable that pain, as we understand it, is entirely wanting in at least many of the insects, the sense of feeling being developed only suffi-ciently to cause them to avoid danger, for I have seen a cruel-minded boy pin a dragon fly to a board and then feed it with numerous house flies, which the dragon fly greedily devoured.

The seventeen-year locust is a Methuselah among insects. It lives seventeen years under ground, where Methuselah did not go until he quit living. But this locust is seldom seen, while the other varieties are with us every summer. The dried shells of the pupæ have been the playthings

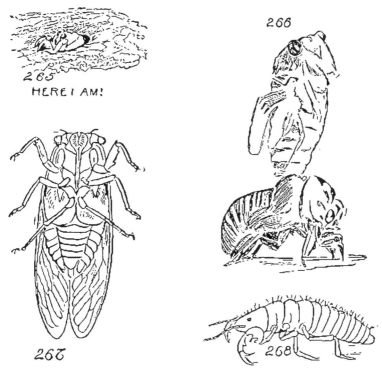

265
HERE I AM!

266

267

268

of children ever since this country was inhabited by white people and no doubt little Indian children played with them before the white people came. Probably the red youngsters sat around and watched the Cicada come out of their hole, as in Fig. 265, to creep up the trunk of a tree, fasten

themselves there with their hooked claws, hunch
up their shoulders until they split the back of the
pupa case, then slowly work their way out until
they looked like ghost-bugs riding on the back of
some queer steed (Fig. 266).

I could not resist the temptation of putting
Fig. 266 with the pupa in a horizontal position,
although that is not the position it assumed while
the Cicada was coming out of the shell. The
ghostly locust itself, at this stage, would be horizontal, that is, parallel with the ground, but the
thing looked so funny standing upright that I
allowed the drawing to be placed in that position.
Fig. 267 shows the under side of the harvest fly or
Cicada and Fig. 268 shows the young Cicada.

Once in Kentucky I went to a dance at Latonia
Springs. It was one of those old-fashioned South-
ern affairs where dancing began at two o'clock
in the afternoon, continued until supper time, and
indefinitely thereafter. But the reason I remember this particular dance is not because of the presence of many beautiful ladies, although there were
assembled there the prettiest girls in the State
noted for its beautiful women, nor is it because of
the fascination of my partner in the dance, although

her graces were many, but it is all due to the fact that this dance happened in the midst of the seventeen-year-locust season! The locusts flew through the ball-room and banged against the men's faces, the ladies knocked them about with their fans, using the latter after the manner tennis players use their racquets. The red-winged bugs were under foot and made the floor more slippery than did the wax with which it was covered, and ever and anon some lady would give a shriek as she suddenly and frantically clutched at her bosom, then she would be hustled into the dressing room by the colored mammy who presided there, and the offending locust removed.

But this was in Kentucky, not only in Kentucky, but in the good old days in Kentucky, and no swarm of seventeen-year locusts was ever hatched that was numerous enough and annoying enough to spoil the fun or seriously interfere with the merriment of a Kentucky picnic dance at that time and place.

The seventeen-year locust, that is, the females, have a sort of ovapositor (egg layer) (Fig. 269) equipped with two so-called saws which are really more like rasp-files. One is on each side, as you

may see from the illustration, and the bug can
work them up and down and thus saw holes in
the green twigs wherein to safely hide her precious
eggs. Fig. 270 is a section of the saw cut cross-
wise, made after a drawing by
Grant Allen, showing how
neatly the parts fit together. Fig.
271 shows the twig with the eggs
in it.

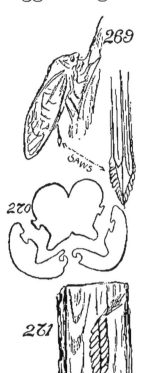

The eggs of the seventeen-year
locust hatch and the baby locust
drops or jumps to the ground, and
then with its powerful claws, digs
until he finds a root, where he stays,
sucking the juices of the roots of
the trees, for seventeen long years,
then he comes out in the sunshine
to sing a while, mate and die.

Some time when you are
afield, you may be lucky enough
to see the big wasp or hornet that feeds its
young with the Cicada, which it captures and
paralyzes with its sting, then lays its eggs upon
it and buries it. This is much better than cold
storage. The young are in no danger of ptomaine

poisoning for the good reason that their meat is not dead, but alive, and it stays alive until they themselves kill it by eating it, which of course happens after they have hatched out of the egg, though generally speaking I suppose I should say after it hatches out of the egg.

This is all interesting, but not half so interesting as watching the Digger wasp lug the poor Cicada over the rough ground, as I have watched it do, to the trunk of a tree, then ascend the tree to its lower branches, dragging the benumbed and paralyzed Cicada after it until the wasp reaches the spot where it can spring into the air and by the aid of its rapidly buzzing wings as a motor, glide slantingly down to the ground again, only to again drag the Cicada to another tree and go through the same process until it reaches the grave it has dug for the poor harvest-fly.

One time in the mountains of Pike County, I heard a Cicada singing " to beat the band." There was nothing particularly remarkable about the musical part, because the dry rasping notes of the Cicadas could be heard in every direction—the trees were full of them. But this one was singing while it was flying and it was flying in a most peculiar manner. In place of winging its way from one tree to another after the custom of its tribe, it darted back and forward, this way and that, over my head, circling and going in spiral in a most erratic style. At last I discovered that a great big cruel dragon fly had captured the poor locust and the locust's song was really a cry for help and that it was not flying at all, but was carried about by its captor.

These are the little incidents, boys, which make the study of insects interesting. It is the life, the habits and the tragedies of the insect world that give us moving-picture stories of adventure which we like to see for ourselves.

CHAPTER NINETEEN

WATER-BUGS

In the outskirts of old Flushing, Long Island, there is an ancient mill-pond where formerly stood a quaint, low-ceilinged, dusty mill dating back to Revolutionary times. Below the mill wheel where the water ran into the brook was formerly a great hunting ground for newts, salamanders and other aquatic animals, but up in the pond itself, in the black soft mud, was our hunting place for all manner of small aquatic bugs.

The mill pond is now dignified by the name of Kissena Lake, and the old mill is gone. There are walks, drives, rustic stairways and caretakers, and the place is called Kissena Park.

But down in the mud of Kissena Lake the little water people still live and thrive. There you will find the Boatman (Fig. 272) not quite half an inch long and he makes an interesting specimen for your aquarium, where he will soon make himself at home and spend his time anchoring himself at the bottom with his middle legs grasping a pebble while his arms are doubled up under his chin and his hind legs all set ready to row like a pair of oars in a shell boat, as indeed they are, not in a boat, but they are oars to propel the boat-bug to the surface when he needs air.

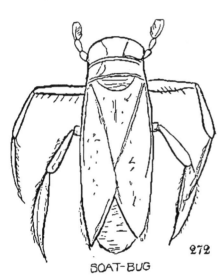

BOAT-BUG

Enlarged View.

The air he takes down with him from the surface in minute silver bubbles, clinging to the outer edge of each upper wing, filling the spaces between the wings and the belly and between his head and chest and sticking to the hairs on his legs like silver spangles.

272

The water Boatman is a great diver and he can stay under water a long, long time without being compelled to come to the surface.

Occasionally these bugs will leave the water and I have found them flying around the kerosene lamps in the farmhouse. Water is water to them, whether it is salt or fresh, and you can find them in the briny lakes of the West and also in the sparkling translucent trout streams.

Down in Mexico the natives collect the eggs of the water boat-bug that inhabits the lakes near the city of Mexico, and according to Mr. Howard they make the eggs into cakes, mixing the eggs with meal before baking them. But here in the United States we do not eat water-bugs' eggs.

BACK-SWIMMERS

Many insects are supplied with many eyes; some of the water beetles have eyes on the top of their heads for looking into the sky and eyes under their heads for looking down into the water. The extra eyes are called OCELLI. The eyes of the back-swimmer are triangular and he has no extra ones scattered about his person. There are several kinds of water Boatmen, but you will find that out when you make your collection.

If you pick up some of these back-swimmers (Fig. 273) with your hands, do not be at all surprised if they jab you with their beak; but you need not be alarmed, don't drop your captive; say " Ouch! " but put him in the pail. Some say the prick from a water Boatman is as painful as a bee sting, but this is not so with me—I have tried them both. The back-swimmer's sting only lasts for an instant and it then is over, but a bee sting hurts and it hurts real good and strong and lasts some time. I never have had the bite of a water Boatman swell up and become inflamed, but a bee, a

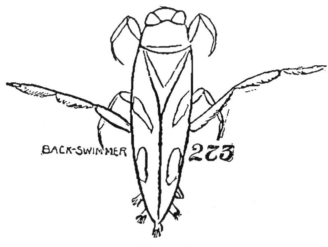

Enlarged view,

yellow-jacket, a hornet or a wasp will raise a great welt on my skin and pain me enough to make me cry if I were not a big man and ashamed to do so childish a thing. But be careful with all waterbugs, as some of them can sting viciously.

There are about twelve species of back-swimmers to be found in the United States and there is no good history of the life of one yet written. So here is a chance for my readers to distinguish themselves.

WATER SCORPIONS

These water-bugs are called scorpions because their front legs, with which they grasp their prey, and their tail combined, give them the appearance of—or rather, suggest, a scorpion. Water scorpions have wings (Fig. 274). The front wing is horny after the manner of bugs and the hind wing is thin, transparent and skinny. They are very flat bugs and like the boat-bugs and the back-swim-

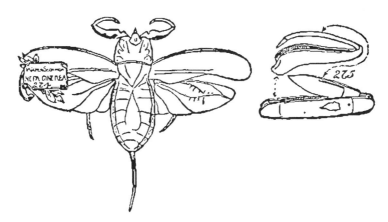

mers they prey upon other water creatures. The water scorpion also has a habit of feeding on fish eggs. It is said to be able to sting severely. Let some scientist try the experiment and accept his report. The report will not pain you.

When you are digging in the mud of Kissena Lake or almost any other pond in our country, you are liable to bring out of the bottom an elongated water-bug known as the Ran´a-tra. He is a long slim fellow with long legs and long horny append-ages at the hind end which it can put together, making an air tube. Both of these water scorpions can shut up the front part of their front legs into the next joint like the blade of a knife into a knife handle (Fig. 275). The Ranatra, one may see at

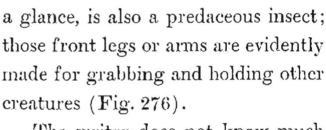

a glance, is also a predaceous insect; those front legs or arms are evidently made for grabbing and holding other creatures (Fig. 276).

The writer does not know much about the personal habits of the Ranatra, and he very much doubts if any other writer has made much of a study of it. The Ranatra does not go skipping about and attracting attention like the Boatmen and the back-swimmers; he looks too much like a stick to be seen, unless one is looking espe-cially for him.

THE GIANT WATER-BUG

These are the big fellows that people call elec-tric-light bugs because they sometimes fly about the electric lights at night. They are the ones that will catch fish in your aquarium as already mentioned in the Fore Talk on pages eight and nine.

The giant water-bugs are homely, forbidding-looking creatures (Fig. 277), and are the biggest bugs in the bug family. They hide in ponds and will catch any small live thing, fish or frog that comes their way, grasping them with those scorpion-like front claws, jabbing them with their beak and probably paralyzing them with the poison spittle which they pour into the wound.

A smaller specimen of a water-bug, built on the lines of the giant one, lived all this last winter in my aquarium, and was plastered all over its shoulders and legs with eggs.

The American observer, Miss Slater, has said that the female bug has a habit of laying her eggs on her husband's back. The old gentleman objects to it most strenuously, but his wife, as the cowboys say, wears the chaps—that is, the leather breeches; in other words, she is master. Miss Slater further says that the gentleman bug, although naturally a lively fellow, feels so disgraced and depressed with his load of eggs that he will not even get out of the way of an enemy, apparently preferring to die than be disgraced by acting the part of a nurse girl.

SKATERS OR GLIDERS

The gliders (Fig. 277½) always interest children; the marvellous way they skate on running water without wetting their feet is an endless source of wonder to the young people.

When the writer was a little fellow he visited a beautiful country in Ohio. There was a brook

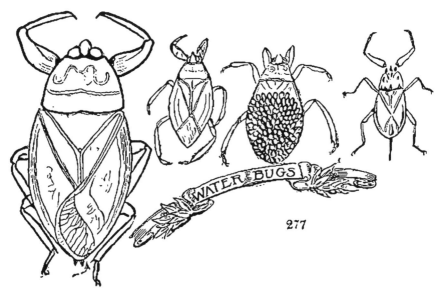

277

about six or seven feet wide, with clear, sparkling water in which one could see the little fish darting around, and over the surface of which the whirligig beetles made spirals and the gliders or skaters, skimmed. The foot of a glider makes a dent in the water, just as if the water had a thin skin on the surface which had been pushed in. This dent

makes a sort of a lens like the lens in a camera or
an opera glass. The sun shining down on the
gliders and on the dents in the water, casts enlarged
shadows on the bottom of the stream and one never
tires of watching these shadows—that is, if one is a
little fellow and has not yet had his mind warped
by business, professional duties, or politics.

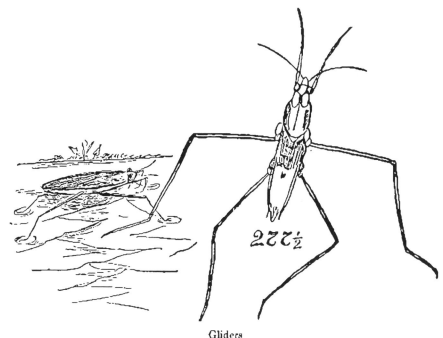

Gliders

The war of the States broke out, the pretty
country place was changed into a busy camp called
Camp Dennison. One of the writer's brothers was
up there as a Union soldier when the little brook
was again visited, but war is the most unnatural
thing, and it and nature cannot agree. The green

sward was gone, the beautiful trees were cut down, the banks of the stream were cut, bruised and torn by the sharp-shod feet of horses. No more little fish could be seen and even the whirligigs and skaters had disappeared, while the stream itself was nothing but liquid mud.

But this happened a long time ago, and there are other streams and a new crop of youngsters to enjoy them. There is a brook on the writer's farm in Connecticut and there the skaters and whirligigs and all the little people of the brook flourish and the author's own little boy and little girl never tire of feeding the gliders with flies and other insects which they catch for them.

The water and its inhabitants are very beautiful and very interesting, but as a rule they seem to be very savage creatures which inhabit the brooks and ponds—even more so, if possible, than those which inhabit the land. The Caddice worms and a few other under-water people live on vegetation, but the rest of them seem to live on each other. Still, they are not parasites nor dead-beats; they belong to the higher order of hunters and fishermen, and the hunting animal or insect must have intelligence in order to succeed.

If you will dip up a few of the gliders with a little net made of cheesecloth and put them in your aquarium, you can tame them and they will learn, like the whirligig, to take the fly from your fingers. But you must keep your aquarium covered with a wire screen or they will escape. Some of them have wings and can fly and all of them will attempt to get away by crawling up the sides of the aquarium. These surface insects seem to dread captivity. The divers and under-water folk do not seem to mind confinement, but all of them will become accustomed to their narrow crystal prison and furnish you a never-ending source of entertainment if you treat them properly.

A FEW MORE BUGS

Somewhere at the fore part of this book I told the reader that there were far too many insects in the United States to squeeze in between the covers of any one book, and any of the bugs who find their portraits missing in this volume will please accept the apologies of the author and the assurance that no slight was intended. There are a few bugs we will mention because the boys will look for them.

There is the squash-bug (Fig. 278), and his near relative, the stink-bug (Fig. 279), also the comical little toad bug (Fig. 280). We have omitted many portraits of the marsh treaders and of the

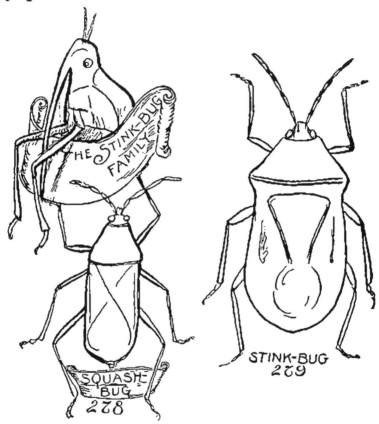

THE STINK-BUG FAMILY

SQUASH-BUG 278

STINK-BUG 279

THE ASSASSIN BUG

notorious bed-bug family, but the less-known assassin bug you will find in Fig. 281.

The assassin belongs in the kissing-bug family. Some years ago there was a great ado in the newspapers about the kissing-bugs stinging people on the mouth and causing their lips to swell up. How much of it was true we do not know, and for the sake of the people said to be kissed by this bug we hope that none of the reports were true, because the bugs accused of promiscuous osculation are the very useful bed-bug hunters; but however useful they are, they are the last of the bug tribe which one would want on one's lips.

My first experience with an assassin bug was one summer day on Long Island, when I was idling away a summer's day, leaning on the paling

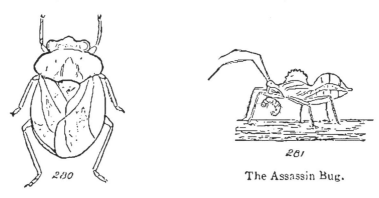

280

The Toad Bug.

281

The Assassin Bug.

fence and talking to my pet red-tailed hawk. While so engaged I noticed the keen-eyed hawk was watching something on the top rail of the fence. Following the direction of its gaze I saw an ugly small-headed creature of the bug family strolling leisurely along the top rail. It did not hurry, but walked as if it had no train to catch. It strolled,

as I have said, leisurely, until it came to a small
caterpillar hurrying by. Just as it stood over the
caterpillar it stopped, and so did the caterpillar,
for the assassin's stilletto pierced the worm-like
body of the baby moth and ended its career right
there.

According to all accounts, both the kissing-bug
and the assassin-bug can make painful wounds, so
it may be well for the young collector not to experiment with them in that line or to allow the ugly
things to poke their sharp beaks through the collector's skin.

CONCLUSION

This book, boys, was written, not to take the
place of any other book in the field, but to stimu-
late your interest and encourage you to read other
books which take up the subject in a more technical
manner—books like "Caterpillars and Their
Moths," which is brim full of original investigation;
but beyond all this and above all this is the hope that
this book will encourage you to go afield and hunt
the insects and study them first hand. Such work
will develop your power of observation.

Boys' eyes are keener than the eyes of men or
grown people. Boys see more, and if their ob-

servation is trained they will learn more than grown people. They will learn to appreciate men like Thoreau and my good friend John Burroughs, men like Dr. Frederick Lucas and Dr. Frank E. Lutz, who give up their lives to the study of nature.

But if you live in the city do not be discouraged, the parks and vacant lots are full of interesting specimens, and after you have learned where to hunt for them you will find them. If this book of Bugs, Butterflies and Beetles really starts you on the road as a student of nature the writer will consider that the book is an unqualified success, for a nature student is one who feels a sympathy, a companionship, for all live creatures however lowly they may be.

Such a feeling broadens the mind and the study sharpens the wits and teaches one how to observe. The pursuit of nature will give you a hobby which will be an interesting and useful pastime, will lighten the cares of business, lengthen your years, take you in the open, where you will gain health and strength, give you good digestion, bright eyes and above all make you happy, cheerful and companionable—in fact it will round out your character so that you will stand high in the estimation of your friends and fellow citizens as does the late John Muir and our patron saint Audubon.

For, after all, it is only those studies and pursuits which make better citizens of us which are worthy of our pursuit, and in closing I want to thank my readers for travelling along with me among the fields and forests, brooks and farms, which made me feel again like a happy twelve-year-old boy.

INDEX

Made in the USA
San Bernardino, CA
30 August 2014